Guidebook to R Graphics
Using Microsoft® Windows

W9-AAZ-087

Guidebook to R Graphics Using Microsoft® Windows

Kunio Takezawa

National Agricultural Research Center,
National Agriculture and Food Research Organization
Graduate School of Life and Environmental Sciences
University of Tsukuba Ibaraki
Tsukuba, Japan

A JOHN WILEY & SONS, INC., PUBLICATION

Library of Congress Cataloging-in-Publication Data:

Takezawa, Kunio, 1959–
  Guidebook to R graphics using Microsoft Windows / Kunio Takezawa.
      p. cm.
  Includes index.
    ISBN 978-1-118-02639-7 (pbk.)
1. Computer graphics. 2. R (Computer program language) 3. Microsoft Windows (Computer file) I. Title.
    T385.T346 2012
    006.6'633—dc23   2011049806

Printed in the United States of America.

10 9 8 7 6 5 4 3 2 1

CONTENTS

Preface   ix
Acknowledgments   xiii

1   Basic Graphics   **1**

1.1   Introduction   1
1.2   Downloading and installation of R   1
1.3   Start-up of R, and construction and execution of R
       programs   3
1.4   Coordinate axes   11
1.5   Points and straight lines   14
1.6   Reuse of graphs produced by R   16
1.7   Text   24
1.8   Various points and straight lines   27
1.9   Fonts   34
1.10   Figures such as circles and rectangles   35
1.11   Legends and logarithmic plots   41
1.12   Bar charts   42
1.13   Pie charts   45

1.14	Layout of multiple graphs	46
1.15	Summary	60
	Exercises	62

2   Graphics for Statistical Analysis   **65**

2.1	Introduction	65
2.2	Stem-and-leaf displays	66
2.3	Histograms and probability density functions	67
2.4	Strip chart	73
2.5	Boxplots	75
2.6	Multiple-axis layouts	80
2.7	Display of confidence intervals	91
2.8	Scatter plot matrices	93
2.9	Radar charts and parallel charts	95
2.10	Functions of one variable	97
2.11	Functions of two variables	100
2.12	Map graphs	108
2.13	Histograms of two variables	113
2.14	Time series graphs of two variables	116
2.15	Implicit functions	119
2.16	Probability density functions	121
2.17	Differential values and values of integrals	124
2.18	Summary	132
	Exercises	133

3   Interactive R Programs   **139**

3.1	Introduction	139
3.2	Positioning by mouse on a graphics window	140
3.3	Inputting values on the console window to draw a graph	143
3.4	Reading data from a data file	156
3.5	Moving data on a natural spline	158
3.6	Understanding simple regression	166
3.7	Adjusting three-dimensional graphs	175
3.8	Constructing polynomial regression equations interactively	180
3.9	Understanding local linear regression	183
3.10	Summary	188
	Exercises	190

4 **Graphics Obtained Using Packages Based on R** **193**

    4.1   Introduction   193

    4.2   Package "rimage"   194

    4.3   Package "gplots"   195

    4.4   Package "ggplot2"   200

    4.5   Package "scatterplot3d"   203

    4.6   Package "rgl"   207

    4.7   Package "misc3d"   221

    4.8   Package "aplpack"   232

    4.9   Package "vegan"   234

    4.10   Package "tripack"   236

    4.11   Package "ade4"   238

    4.12   Package "vioplot"   241

    4.13   Package "plotrix"   243

    4.14   Package "rworldmap"   247

          Exercises   249

5 **Appendix** **253**

    A.1   Digital files   253

    A.2   Free software   254

    A.3   Data   254

Index   257

PREFACE

Carol Marcus: Let me show you something that will make you feel young as when the world was new. (Star Trek: The Wrath of Khan (1982) http://www.imdb.com/title/tt0084726/quotes)

Construction of appropriate graphs plays an important role in data analysis. Pertinent graphs often reveal the conditional implications of data clearly even if the summarization of data by deriving a small number of values can show limited aspects of the characteristics of data. In addition, persuasive graphs are an indispensable tool for the presentation of scientific papers, the description of commodities, patent applications, project proposals, lectures, training courses, business meetings, negotiations, legal actions, etc. The graphical presentation of features of data and structures of concepts can strengthen arguments.

Therefore, it is well known that producing high-quality graphs is a requisite in diverse fields. Demand for richly expressive graphs has grown. Moreover, people's need for better graphs at a lower cost has increased markedly because the extraordinary development of computer technology has enabled the drawing of complicated graphs in a short period of time. Therefore, drawing graphs using a PC is no longer a specialist skill that requires professional expertise. It should be a fairly commonplace technique, comparable to the creation of simple documents using a word processor.

The free software "R" is now widely used for statistical calculation and graphics. R is equipped with various functions for constructing graphs. Moreover, installing software packages enables a wider range of graphs to be produced. The importance of R as a tool for graphics has increased. However, those unfamiliar with the use of R for statistical calculations tend to avoid the software because its wide range of functions may make it appear daunting.

In light of this situation, this book aims to demonstrate that producing graphs using R is an easy-to-master technique. Hence, this book often does not describe R commands and their range of arguments exhaustively, but rather exemplifies typical methods for constructing graphs and their results. This makes the book a guide that aims to foster a feeling of confidence that most graphs can be produced using R. As is generally the case with learning word processing software, learners should obtain an overall picture of the graphical abilities of R in a short time by gaining familiarity with its main features rather than by acquiring details of each function; then they can realize a greater diversity of graphics by referring to references and articles on the Internet. This way of learning is effective provided learners are not aspiring to careers as specialists in the software.

To achieve the aims of this book, procedures for excuting R are presented as R programs that contain a series of R commands. That is, this book, unlike other books in this field, does not proceed with procedures by inputting R commands sequentially. Batch processing is focused on rather than sequential processing (real-time processing). Sequential processing has the advantage of being able to construct graphics while seeing the graphs in progress. However, batch processing is clearly superior to sequential processing because inappropriate manipulation and keystroke errors often occur during the work. In batch processing, R programs (a series of R commands) are recorded regularly, making it easy to rerun a procedure after modifying the program. In addition, learners can review R programs that they have learned, and use or develop them after accumulating techniques in the form of R programs.

Fortunately, if learners transform their R programs into text files, the programs can be easily found when needed by full-text searching. For example, we assume that a folder called **D:\GraphicsR** contains **.RData** (a work image file; this file stores R programs and data). When R programs are accumulated in **.RData**, it will be useful for learners to search for R programs that they need among the programs in **.RData**. For this purpose, learners should activate **.RData** and run the R program below, for instance.

```
function()
{
  ob1 <- objects(pos=1)
  nd <- length(ob1)
  for(ii in 1:nd) {
    print(paste("D:\\GraphicsR\\", ob1[ii], ".txt", sep = ""))
    dump(ob1[ii], paste("D:\\GraphicsR\\", ob1[ii], ".txt",
```

```
      sep = ""))
   }
}
```

This procedure transforms all objects (R programs and data files) in `.RData` into text files and outputs these text files in `D:\GraphicsR`. Extensions of these text files will be `.txt`. If a full-text search of these text files is carried out with R commands as key words, the R programs that use these R commands will be found. Additionally, if comments such as "draw a circle" or "change a font" are written in R programs, a full-text search with the key words of "circle" or "font" will find R programs that contain the above functions.

Moreover, this book explains the method of producing GUI (interactive) programs. Although many people need to construct graphs or carry out statistical calculations, some would rather not go to the trouble of inputting R commands sequentially based on a process of trial and error while considering the function of each R command, and they would have little ability to cope with R programs consisting of several tens of lines. To allow such people to benefit from the useful functions of R, interactive R programs should be prepared to create an environment in which R can be used free from the consciousness of its presence. This is a goal of this sort of program. In addition, if interactive R programs are used in lectures on statistical analysis or as exercise materials for students taking such a course, the contents of statistical analysis can be conceived as a smooth flow of concepts and statistical analysis will attract a high level of interest. The author hopes that a great number of readers will appreciate this intention to construct R programs along this line to increase the familiarity with R.

For these purposes, the R programs listed in this book are elementary ones that allows readers to understand each function of R. New R programs can be developed by adapting these R programs according to readers' needs and interests. By transforming them into text files by the method described above, these text files will guide readers in producing graphs using R. This book provides a platform for readers' originality and ingenuity in this manner; it is not an introductory book that must be read in page order. Such constructive usage of this book is expected to expand the possibilities for creating graphics using R programs, regardless of whether this book is read alone, used as a reference, or used as a textbook in a course on statistical analysis.

The FTP site for this book can be found at the following URL: `ftp://ftp.wiley.com/public/sci_tech_med/guidebook_r-graphics`.

The R programs listed in this book assume that the size of the graphics window on the display of your PC is roughly 16 cm (roughly 6.3 inch) × 16 cm (please refer to the figure below). If a larger graphics window than this is needed, or the graphics window cannot be as large as this because of the limited size of your display, the R programs should be modified.

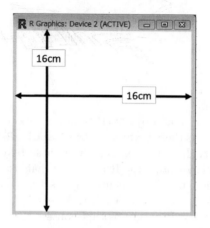

Figure 0.1 Graphics window

Warning: When the R programs listed in this book are executed or modified by trainees in a lecture or a workshop, the copyright holder has made it a condition that all trainees have purchased this book.

KUNIO TAKEZAWA

Tsukuba, Japan
March, 2012

ACKNOWLEDGMENTS

The original edition of this book was published as "Graphics and GUI Operations Using R" ("R niyoru gazouhyougen to GUI sousa"), written in Japanese. The publisher is CUTT System Development Laboratory Inc. (Hyakunin-cho 4-9-7 8th floor, Shinjuku, Tokyo 169-0073, Japan. President: Mr. Katsutoshi Ishizuka, Editor: Mr. Tomohiro Takei). CUTT System Development Laboratory Inc. kindly agreed to our request to publish an English edition of this book. This English edition was not just a faithful translation, but underwent processes of additions and alterations on the advice of the editors at Wiley.

To all these wonderful people I owe a deep sense of gratitude, especially now that this project has been completed.

K.T.

CHAPTER 1

BASIC GRAPHICS

1.1 INTRODUCTION

This chapter first describes the procedures to start up R, produce R programs, and run them. This is followed by explanations of R programs used to construct simple graphs. New techniques are introduced by adding new methods to already-known materials. Hence, learning is followed by consulting, testing, and modifying the listed R programs sequentially. Alternatively, learners can find graphs that roughly suit their purpose and experiment with them. Previous articles may be referred to if unknown commands or functions are used in the programs. In addition, the last part of this chapter introduces techniques to share displayed graphs with other application software packages and to save graphs as digital files.

1.2 DOWNLOADING AND INSTALLATION OF R

The procedure below installs R on a PC loaded with a Windows OS.

Guidebook to R Graphics Using Microsoft Windows,
First Edition. By Kunio Takezawa
Copyright © 2012 John Wiley & Sons, Inc.

1. Access "the R Project for Statistical Computing" web page (http://www.r-project.org/).

2. Click "CRAN" under "Download, Packages" listed in the menu on the left side.

3. Choose one of the mirror sites on the "CRAN Mirrors" page. Most of the mirror sites are identical. For example, choose University of Tsukuba (http://cran.md.tsukuba.ac.jp/).

4. Click "Windows" under "Download and Install R" on "The Comprehensive R Archive Network" web page.

5. Click "base" on the same line as "base Binaries for base distribution (managed by Duncan Murdoch)" on the "R for Windows" page.

6. Click "Download R 2.11.0 for Windows (32 megabytes)" (the version may be different) on the "R-2.11.0 for Windows" page.

7. R-2.11.0-win32.exe (or a different version) is now available for installation. Construct a folder in the hard disk in your PC and download the file to the folder.

8. Double-click R-2.11.0-win32.exe to start the installation process of R.

9. Agree to the "GNU General public license" and specify the location of installation.

10. Select "Full installation" on the "Select components" page.

11. After a few further selections, installation starts.

Sets of programs called "packages" have been prepared for R. Packages can be added to the version of R installed on your PC. If a PC is connected to the Internet, select "Packages" in a menu after R is booted. Then, choose "Set CRAN mirror...". Many mirror sites appear. Then, choose "Japan (Tsukuba)", for example, and click "OK". Return to "Packages" in the menu. Then, select "Load package" to display the names of many packages. Choose the names of packages that you need and click "OK". The selected packages are installed in this way.

If packages are installed in this manner, the names of packages displayed are those that fit the version of R installed on a PC. However, some packages that fit an older version of R can be used in a newer version of R. For example, the "gtools" package that will be dealt with later fits version 2.10 of R but does not fit version 2.11 of R. Hence, when the "gtools" package is used, it is safe to employ version 2.10 of R. However, the use of the "gtools" package with version 2.11 of R can be attempted. For this purpose, the method described above cannot be used. A method for a PC that is not connected to the

Internet is useful for installing packages for older versions of R. This method is as follows:

1. In the fifth step of the installation procedure of R described above, click "contrib" on the same line of "contrib Binaries of contributed packages (managed by Uwe Ligges)".

2. The "Index of /bin/windows/contrib" page appears. Select a version of R that fits the packages that you will use (for example, 2.10).

3. The names of the files of packages compressed in zip format are displayed. Click the names of the files of packages needed for downloading them.

4. R is booted. Select "Packages" in the menu. Then, select "Install package(s) from local zip files...".

5. Select the files that were downloaded beforehand to install packages.

Using this method, even if R version 2.11 is installed on a PC, packages for R version 2.10 may be used.

1.3 START-UP OF R, AND CONSTRUCTION AND EXECUTION OF R PROGRAMS

We assume that the folder used for storing data files and the results of calculations is D:\GraphicsR. Save all files created with R in this folder.

.RData

Figure 1.1   Work image file.

The file for storing R programs is called the "work image file" (Fig. 1.1). Confirm the presence of this file in D:\GraphicsR. If the work image file is not located in D:\GraphicsR, search for a file named ".RData" (which will be somewhere in your hard disk if R has been installed correctly) and copy the file by saving it in D:\GraphicsR. If the drive where .RData is originally located is the "D" drive, hold the Ctrl key and copy the file. If the Ctrl key is not held, the file is not copied but moved. That is, the original .RData is deleted. When multiple .RData files are placed in the same PC, each .RData

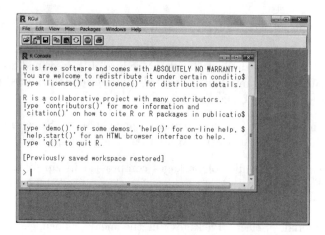

Figure 1.2   Window immediately after "R" is booted.

file stores its own R programs. Even if the version of R is upgraded, the same .RData file can be used.

R is booted by double-clicking .RData in D:\GraphicsR. R can also be booted by double-clicking the shortcut button assigned to .RData on the desktop or other places. Upon booting R, the window shown in Fig. 1.2 appears.

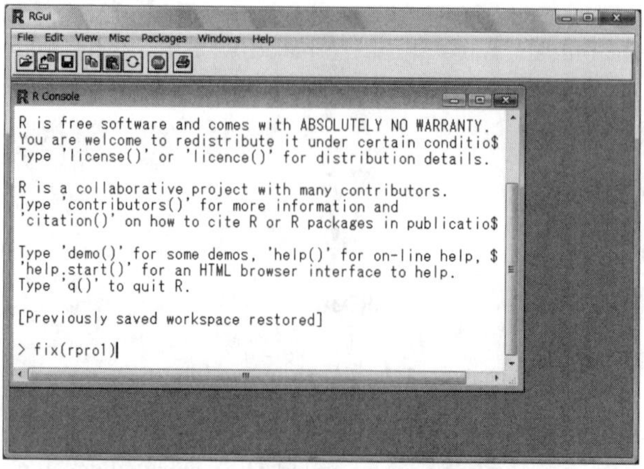

Figure 1.3   Construction of R program.

The inner window in Fig. 1.2 is called the console window. In this window, R commands and R programs are executed. For example, when an R program named rpro1() is produced, type fix(rpro1) and click the return key in the console window (Fig. 1.3). Then, the display in Fig. 1.4 appears. The new window is an editor. When R is installed in the standard manner, "notepad" is used as an editor. However, users can set up another editor for this purpose. An R program is a series of R commands listed between "`function () {`" and "`}`". Arguments can be specified in "`()`" of "`function () {`". For example, `function (aa) {` is an R program that uses `aa` as an argument. By setting numerical values or text as `aa`, an R program with this argument is run.

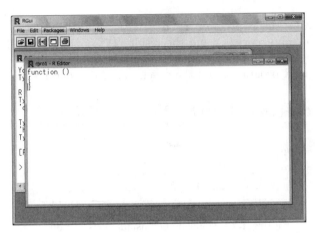

Figure 1.4   Editor.

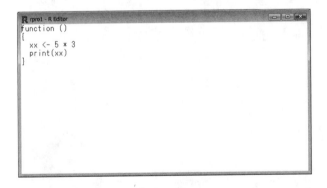

Figure 1.5   R program.

Let us produce an R program for multiplying 5 and 3 and displaying the result in the console window. Fig. 1.5 shows an example of an R program for this purpose. The editor is closed to execute this R program. For this, "×" (Fig. 1.6) located of the upper right of the editor is clicked. Then, a dialogue box asking for a selection is displayed (Fig. 1.7). "Yes" or the return key is clicked. Only the console window is then displayed (Fig. 1.8). `rpro1()` is

Figure 1.6   Symbol clicked to close.

typed to execute the R program **rpro1()** (Fig. 1.9). Then, the return key
is clicked to execute **rpro1()** and the result "15" appears (Fig. 1.10). This
series of procedures summarizes the basic use of R: the start-up of "R", the
construction of R programs, and the execution of R programs.

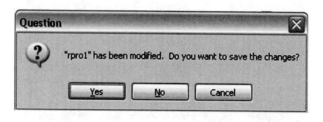

Figure 1.7   Dialogue box asking for selection.

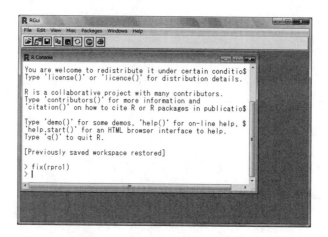

Figure 1.8   Return to the console window.

When graphics windows are displayed in the console window (Fig. 1.11),
the execution of **graphics.off()** in the console window (Fig. 1.12) clears
all graphics windows, and the console window remains (Fig. 1.13). If R is
shut down and rebooted, only the console window appears. Even if only the

```
R R Console                                        [_][□][✕]
You are welcome to redistribute it under certain conditio$
Type 'license()' or 'licence()' for distribution details.

R is a collaborative project with many contributors.
Type 'contributors()' for more information and
'citation()' on how to cite R or R packages in publicatio$

Type 'demo()' for some demos, 'help()' for on-line help, $
'help.start()' for an HTML browser interface to help.
Type 'q()' to quit R.

[Previously saved workspace restored]

> fix(rpro1)
> rpro1()|
```

Figure 1.9   Execution of an R program.

```
R R Console                                        [_][□][✕]

R is a collaborative project with many contributors.
Type 'contributors()' for more information and
'citation()' on how to cite R or R packages in publicatio$

Type 'demo()' for some demos, 'help()' for on-line help, $
'help.start()' for an HTML browser interface to help.
Type 'q()' to quit R.

[Previously saved workspace restored]

> fix(rpro1)
> rpro1()
[1] 15
> |
```

Figure 1.10   Result of executing an R program.

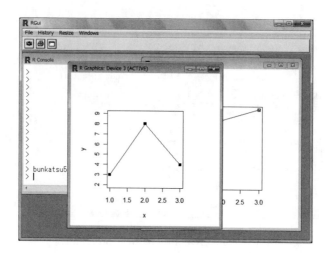

Figure 1.11   Graphics windows are displayed in addition to the console window.

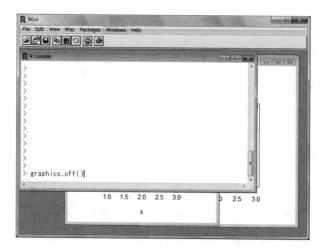

Figure 1.12   `graphics.off()` is executed in the console window.

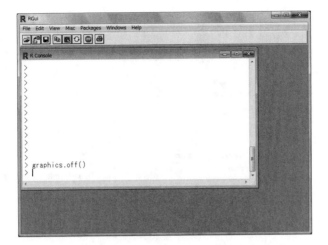

Figure 1.13   The operation shown in Fig. 1.12 clears all graphics windows.

console window is displayed, the configurations of previous graphs may affect new graphs. The shutdown and rebooting of R solves this problem.

R uses Notepad as the standard editor by default. Other editors, however, can be used. For example, "Programmer's Notepad" (`http://www.pnotepad.org/`) is available as an editor: "Programmer's Notepad" has various useful functions. For instance, the setting of "View" - "Change Scheme" - "C / C++" (Fig. 1.14) gives the edit screen such as Fig. 1.15. The structure of the parentheses is shown clearly.

The following R program (`fixp()`) is useful for employing "Programmer's Notepad" as an editor:

```
Program (1 - 1)
function (x, ...)
{
  subx <- substitute(x)
    if (is.name(subx))
      subx <- deparse(subx)
    if (!is.character(subx) || length(subx) != 1)
      stop("'fix' requires a name")
    parent <- parent.frame()
    if (exists(subx, envir = parent, inherits = TRUE))
    x <- edit(get(subx, envir = parent), title = "temp",
      editor="C:\\Program Files\\Programmer's Notepad\\pn.exe", ...)
    else {
      x <- edit(function() {
      }, title = subx, ...)
      environment(x) <- .GlobalEnv
```

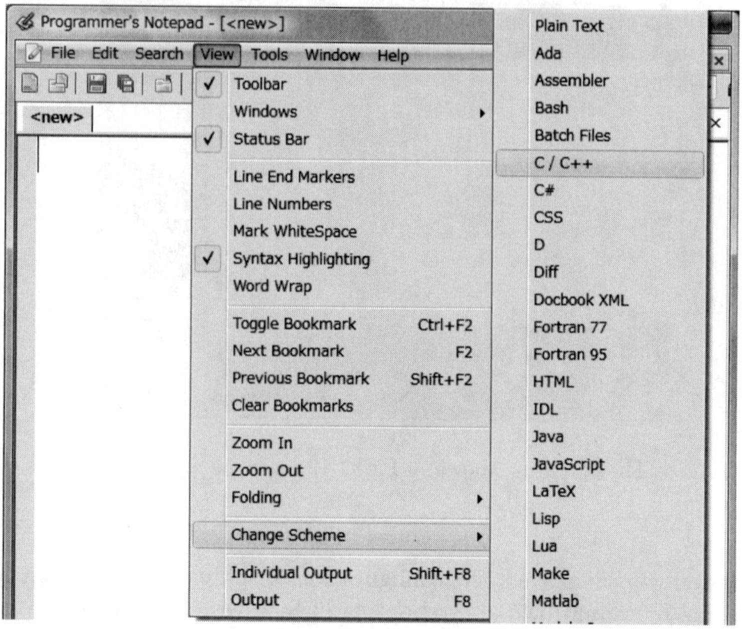

Figure 1.14   Setting of "View" - "Change Scheme" - "C / C++" in "Programmer's Notepad".

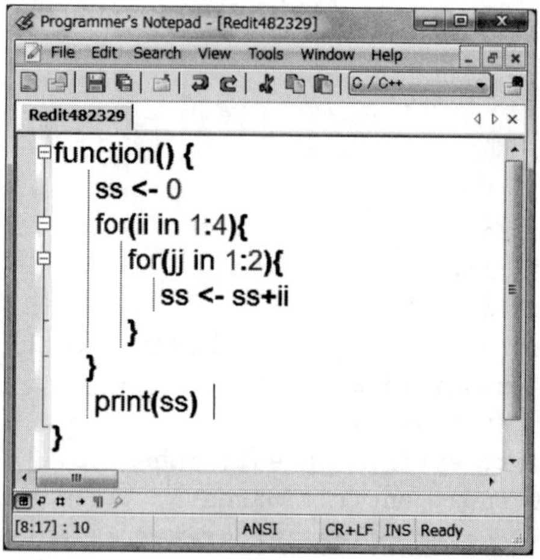

Figure 1.15   Edit screen given by setting Fig. 1.14.

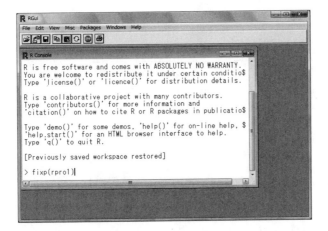

Figure 1.16   Construction of R program using "Programmer's Notepad".

```
  }
  assign(subx, x, envir = .GlobalEnv)
}
```

This R program is obtained by modifying `fix()`, which is implemented in R by default. When an R program is edited using `fixp()`, `fixp()` is carried out on a console window in the same manner as that of using `fix()`(Fig. 1.16).

1.4 COORDINATE AXES

Fig. 1.17 shows the result of Program (1 - 2).

```
Program (1 - 2)
function() {
# (1)
  par(mai = c(1, 1, 1, 1), omi = c(0, 0, 0, 0))
# (2)
  plot(x = c(0, 1), y = c(0, 1), xlab = "x", ylab = "y")
}
```

(1) `par()` sets the structure of R graphics and outputs the setting. In this example, a graphics area is set in a graphics window. The meanings of `omi =` and `mai =` are illustrated in Fig. 1.18. `omi =` sets the size of the outer margin surrounding the graphics area. The four values from left to right indicate the sizes of the lower, left, upper, and right margins of the graphics area (the outer rectangle in Fig. 1.18). `mai =` sets the size of the figure margin. The four values from left to right indicate the sizes of the lower, left, upper, and right margins of the figure area (the inner rectangle in Fig. 1.18). When values or

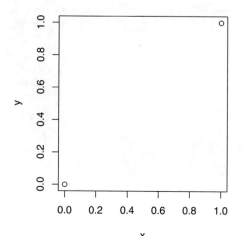

Figure 1.17   Coordinate axes and two data points given by Program (1 - 2).

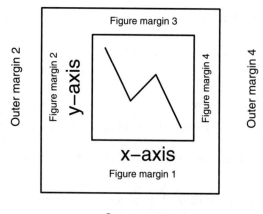

Figure 1.18   Structure of a graphics window.

labels are written on the lower or left side of a line chart or scatter plot, the first two values of `mai =` are usually set to be larger than the last two values. The unit for `omi =` and `mai =` is inches. However, when graphs are drawn on a display, these values indicate relative lengths because the sizes of graphs can be changed at will.

(2) Fig. 1.17 shows that `plot()` illustrates a graph with two data points at $(0,0)$ and $(1,1)$. Furthermore, `xlab = "x"` adds the label "x " on the x-axis of the graph and `ylab = "y"` adds the label "y " on the y-axis of the graph.

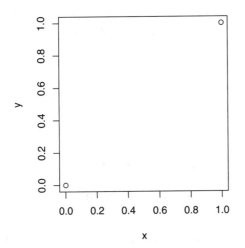

Figure 1.19   Coordinate axes and two data points drawn by Program (1 - 3).

Fig. 1.19 is identical to Fig. 1.17 and is constructed by Program (1 - 3) .

Program (1 - 3)
```
function() {
# (1)
    par(mai = c(1, 1, 1, 1), omi = c(0, 0, 0, 0))
# (2)
    plot(c(0, 1), c(0, 1), xlab = "x", ylab = "y")
}
```

In part (2), the first two arguments of `plot()` (x = c(0, 1), y = c(0, 1),) are replaced with c(0, 1), c(0, 1),. In R commands, the names of arguments can usually be abbreviated if arguments are given in a preset order.

Program (1 - 4) produces Fig. 1.20 (left).

Program (1 - 4)
```
function() {
# (1)
```

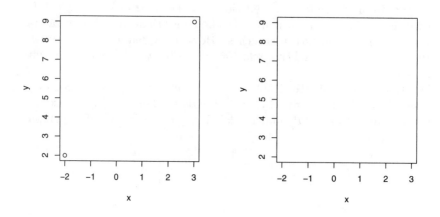

Figure 1.20   Coordinate axes and two data points drawn by Program (1 - 4) (left). Coordinate axes and two data points drawn by Program (1 - 5) (right).

```
  par(mai = c(1, 1, 1, 1), omi = c(0, 0, 0, 0))
# (2)
  plot(c(-2, 3), c(2, 9), xlab = "x", ylab = "y")
}
```

The first two arguments of part (2) are replaced with c(-2, 3), c(2, 9). This alteration changes the positions of the two points to $(-2, 2)$ and $(3, 9)$.
   Program (1 - 5) constructs Fig. 1.20 (right).

Program (1 - 5)
```
function() {
# (1)
  par(mai = c(1, 1, 1, 1), omi = c(0, 0, 0, 0))
# (2)
  plot(c(-2, 3), c(2, 9), xlab = "x", ylab = "y", type = "n")
}
```

In part (2) in Program (1 - 4), type = "n" is added to the arguments of plot(). This removes the two data points so that only the coordinate axes remain. This technique draws the coordinate axes only. The ranges of the coordinate axes are set by the first two arguments.

1.5   POINTS AND STRAIGHT LINES

Program (1 - 6) produces Fig. 1.21 (left).

Program (1 - 6)

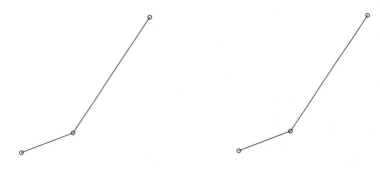

Figure 1.21 Straight line and data points drawn by Program (1 - 6) (left). Straight line and data points drawn by Program (1 - 7) (right).

```
function() {
# (1)
  par(mai = c(1, 1, 1, 1), omi = c(0, 0, 0, 0))
# (2)
  plot(c(-2, 0, 3), c(2, 3, 9), xlab = "", ylab = "", type = "n",
    axes = F)
# (3)
  lines(c(-2, 0, 3), c(2, 3, 9))
# (4)
  points(c(-2, 0, 3), c(2, 3, 9))
}
```

(1) The graphics area is set.
(2) Three points, $(-2, 2)$, $(0, 3)$, and $(3, 9)$, are set by the first two arguments in plot(). Similarly, the first two arguments in plot() can specify more than three points. The coordinate axes are set to position all the specified points appropriately in the graph. The remaining specifications of the arguments in plot() are set to not label the axes, the positions specified by the first two arguments, or the coordinate axes.
(3) Straight lines connecting the three points $((-2, 2)$, $(0, 3)$, $(3, 9))$ are sequentially drawn.
(4) Small circles at $(-2, 2)$, $(0, 3)$, and $(3, 9)$ are drawn.
   Program (1 - 7) constructs Fig. 1.21 (right). The resultant graph is identical to that in Fig. 1.21 (left).

Program (1 - 7)
```
function() {
```

```
#(1)
  par(mai = c(1, 1, 1, 1), omi = c(0, 0, 0, 0))
# (2)
  xx <- c(-2, 0, 3)
  yy <- c(2, 3, 9)
# (3)
  plot(xx, yy, xlab = "", ylab = "", type = "n", axes = F)
# (4)
  lines(xx, yy)
# (5)
  points(xx, yy)
}
```

This R program differs from Program (1 - 6) in that c(-2, 0, 3) is stored in xx in (2) and c(2, 3, 9) is stored in yy. xx and yy are variables (called "objects" in the terminology of R). Parts (3)(4)(5) use xx and yy as arguments. This R program produces the same graph as that in Fig. 1.21 (left) by specifying three data points simultaneously.

1.6   REUSE OF GRAPHS PRODUCED BY R

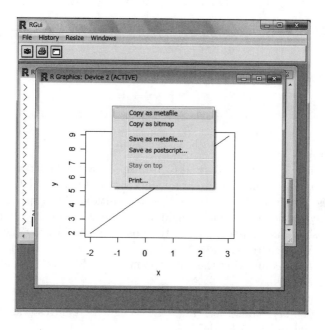

Figure 1.22   Copy of a graph as a metafile.

Undo	Ctrl+Z
Cut	Ctrl+X
Copy	Ctrl+C
Paste	Ctrl+V
Clear	Delete
Paste and Discard	
Copy As	▸
Save Selection As...	▸
Select All	Ctrl+A
Insert Object...	
Motion	▸
Expression Input	▸
Make 2D	Shift+Ctrl+Y
Check Balance	Shift+Ctrl+B
Preferences...	

Figure 1.23   Copy of the metafile in a ;//word processor document.

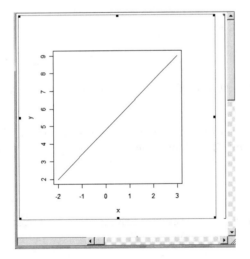

Figure 1.24   Completion of the copy of the metafile in a word processor document.

The graphs produced by R can be arranged using other programs of a personal computer or pasted in a document. For pasting graphs in a graphics window in a document of other programs, "Copy as metafile" is used. That is, a right mouse button is used click on the graphics window and "Copy as metafile" is selected with a left-click (Fig. 1.22). Then, "copy" is performed on a targeted document of the software (a word processor in this example (Fig. 1.23). Now, the paste of the graph constructed by R is completed (Fig. 1.24).

However, if a graph on the graphics window is copied directly, it is cumbersome to unify the lengths or ratios of the vertical axes to the horizontal axes, or the fonts. In this regard, it is useful to save the graphs produced by R and retrieve them as occasion arises. There are diverse formats of digital files available. Postscript files, jpeg (Joint Photographic Experts Group), and pdf (Portable Document Format) are typical examples.

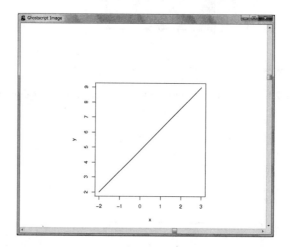

Figure 1.25   Display of a postscript file given by Program (1 - 8).

To construct digital files of graphs in postscript format, Program (1 - 8), for example, is useful. The digital file **ps1.ps** is displayed in Fig. 1.25.

```
Program (1 - 8)
function() {
# (1)
  postscript(file = "d:\\GraphicsR\\ps1.ps", horiz = F,
  width = 5, height = 5)
# (2)
  par(mai = c(1, 1, 0.1, 0.1), omi = c(0.2, 0.2, 0.2, 0.2))
# (3)
  plot(c(-2, 3), c(2, 9), xlab = "x", ylab = "y", type = "n")
  lines(c(-2, 3), c(2, 9))
```

```
# (4)
  graphics.off()
}
```

(1) The command `postscript()` describes that digital files in postscript format is produced. The file name is set as `ssps1.ps`. The folder for this file is `d:\\GraphicsR`. If `horiz = F` is not specified, resultant graphs are rotated by 90 degrees. Hence, this setting is requisite. The arguments `width = 5` and `height = 5` set the size of the image. The unit is inches.
(2) The graphics area is set.
(3) A graph is drawn.
(4) The task of producing a digital file terminates.

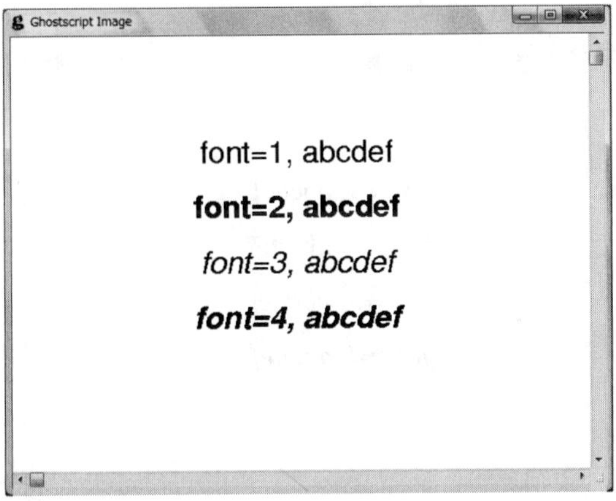

Figure 1.26   Display of a postscript file, given by Program (1 - 9).

When text is contained in a graph to be saved as a postscript file, the setting may be different from that for display in a graphics window. For example, Program (1 - 9) yields a postscript file shown in Fig. 1.26.

```
Program (1 - 9)
function() {
# (1)
  postscript(file = "d:\\GraphicsR\\ps2.ps", horiz = F)
# (2)
  par(mai = c(1, 1, 1, 1), omi = c(6, 0, 0, 2))
# (3)
  plot(c(0, 1), c(0, 1), xlab = "", ylab = "", type = "n",
    axes = F)
# (4)
  text(0.5, 0.8, "font=1, abcdef", cex = 2, font = 1)
  text(0.5, 0.6, "font=2, abcdef", cex = 2, font = 2)
```

```
  text(0.5, 0.4, "font=3, abcdef", cex = 2, font = 3)
  text(0.5, 0.2, "font=4, abcdef", cex = 2, font = 4)
# (5)
  graphics.off()
}
```

(1) The command `postscript()` describes the production of a digital file in postscript format.
(2) The graphics area is set.
(3) The coordinate axes are set.
(4) Text is written. Since the name of the font is not specified in (1), Helvetica is employed. However, another font may be used if the OS setting is different.
(5) The task of producing a digital file terminates.

Figure 1.27   Display of a postscript file, given by Program (1 - 10).

The name of the font can be specified when a postscript file is constructed. The digital file `ps2b.ps` (in postscript format) produced using this method realizes Fig. 1.27 given by Program (1 - 10).

```
Program (1 - 10)
function() {
# (1)
  postscript(file = "d:\\GraphicsR\\ps2b.ps", horiz = F,
    family = "Times")
# (2)
  par(mai = c(1, 1, 1, 1), omi = c(6, 0, 0, 2))
# (3)
  plot(c(0, 1), c(0, 1), xlab = "", ylab = "", type = "n",
    axes = F)
# (4)
```

```
    text(0.5, 0.8, "font=1, abcdef", cex = 2, font = 1)
    text(0.5, 0.6, "font=2, abcdef", cex = 2, font = 2)
    text(0.5, 0.4, "font=3, abcdef", cex = 2, font = 3)
    text(0.5, 0.2, "font=4, abcdef", cex = 2, font = 4)
# (5)
 graphics.off()
}
```

(1) The command `postscript()` describes the production of a digital file in postscript format. The command `family = "Times"` specifies Times as the font here. The font Times is not available if the OS setting is different.
(2) The graphics area is set.
(3) The coordinate axes are set.
(4) Text is written. Since `family = "Times"` is set in (1), Times is employed as the font.
(5) The task of producing a digital file terminates.

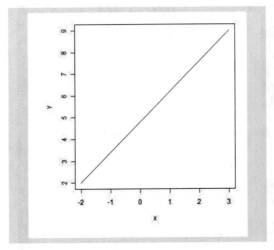

Figure 1.28   Display of a jpeg file, given by Program (1 - 11).

As a format of a digital file for representing graphs, the jpeg format is also available. For example, Program (1 - 11) produces a jpeg file shown in Fig. 1.28.

Program (1 - 11)
```
function() {
# (1)
   jpeg(file = "d:\\GraphicsR\\jpeg1.jpg", width = 400,
     height = 400)
# (2)
```

```
  par(mai = c(1, 1, 0.1, 0.1), omi = c(0.2, 0.2, 0.2, 0.2))
# (3)
  plot(c(-2, 3), c(2, 9), xlab = "x", ylab = "y", type = "n")
# (3)
  lines(c(-2, 3), c(2, 9))
# (4)
  graphics.off()
}
```

(1) The command `jpeg()` describes the production of a digital file in jpeg format. The name of the file is set as `jpeg1.ps`. The name of the folder for saving the file is set as `d:\\GraphicsR`. The arguments `width = 400` and `height = 400` set the size of the image. The unit is pixel.

(2) The graphics area is set.

(3) A graph is drawn.

(4) The task of producing a digital file terminates.

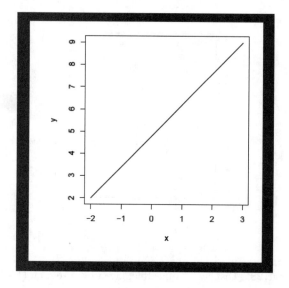

Figure 1.29   Display of the pdf file given by Program (1 - 12).

As a digital file for representing graphs, the pdf format is also available. For example, Program (1 - 12) produces a pdf file shown in Fig. 1.29.

```
Program (1 - 12)
function() {
# (1)
  pdf(file = "d:\\GraphicsR\\pdf1.pdf", width = 5, height = 5)
# (2)
  par(mai = c(1, 1, 0.1, 0.1), omi = c(0.2, 0.2, 0.2, 0.2))
```

```
# (3)
  plot(c(-2, 3), c(2, 9), xlab = "x", ylab = "y", type = "n")
  lines(c(-2, 3), c(2, 9))
# (4)
  graphics.off()
}
```

(1) The command **pdf()** describes the production of a digital file in pdf format. The name of the file is set as **pdf1.ps**. The name of the folder for saving the file is set as **d:\\GraphicsR**. The arguments **width = 5** and **height = 5** set the size of the image. The unit is inches.
(2) The graphics area is set.
(3) A graph is drawn.
(4) The task of producing a digital file terminates.

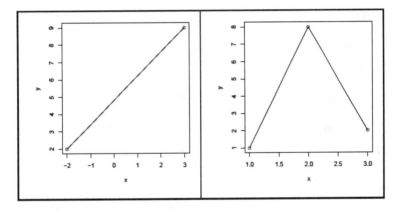

Figure 1.30   Display of the pdf file constructed by Program (1 - 13).

One pdf file may consist of plural pages, each of which contains graphs. For example, Program (1 - 13) produces a pdf file shown in Fig. 1.30.

```
Program (1 - 13)
function() {
# (1)
  pdf(file = "d:\\GraphicsR\\pdf2.pdf", width = 5, height = 5)
# (2)
  par(mai = c(1, 1, 0.1, 0.1), omi = c(0.2, 0.2, 0.2, 0.2))
# (3)
  plot(c(-2, 3), c(2, 9), xlab = "x", ylab = "y")
  lines(c(-2, 3), c(2, 9))
# (4)
  par(mai=c(1, 1, 0.1, 0.1), omi=c(0.2, 0.2, 0.2, 0.2))
# (5)
```

```
plot(c(1, 2, 3), c(1, 8, 2), xlab = "x", ylab = "y")
lines(c(1, 2, 3), c(1, 8, 2) )
# (6)
graphics.off()
}
```

(1) The command **pdf()** describes the production of a digital file in pdf format.
(2) The graphics area for the first graph is set.
(3) The graph for the first page is drawn.
(4) The graphics area for the second graph is set.
(5) The graph for the second page is drawn.
(6) The task of producing a digital file terminates.

1.7 TEXT

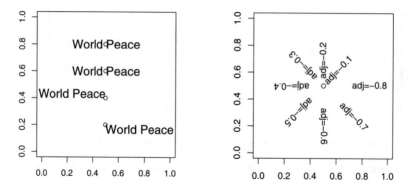

Figure 1.31   Coordinate axes, data points, and text drawn by Program (1 - 14) (left). Coordinate axes and text arranged in a circular pattern drawn by Program (1 - 15) (right).

Text as well as straight lines and small circles can be written in the graphics area defined by **plot()**. Fig. 1.31 (left), produced by Program (1 - 14), shows an example of this.

```
Program (1 - 14)
function() {
# (1)
  par(mai = c(1, 1, 1, 1), omi = c(0, 0, 0, 0))
# (2)
  plot(c(0, 1), c(0, 1), xlab = "", ylab = "", type = "n")
# (3)
```

```
  points(0.5, 0.8)
  text(0.5, 0.8, "World Peace", cex = 2)
# (4)
  points(0.5, 0.6)
  text(0.5, 0.6, "World Peace", cex = 2, adj = c(0.5, 0.5))
# (5)
  points(0.5, 0.4)
  text(0.5, 0.4, "World Peace", cex = 2, adj = c(1, 0))
# (6)
  points(0.5, 0.2)
  text(0.5, 0.2, "World Peace", cex = 2, adj = c(0, 1))
}
```

(1) The graphics area is set.
(2) The coordinate axes are set and drawn.
(3) A small circle is drawn at $(0.5, 0.8)$. "World Peace" is written at the same place. cex = 2 specifies the size of the letters. A comparison between the position of the small circle and that of the text shows that the coordinates in text() indicate the position of the text.
(4) A small circle is drawn at $(0.5, 0.6)$, and "World Peace" is written at the same place. adj = c(0.5, 0.5) is set in text(). However, adj = c(0.5, 0.5) does not affect the position of the text.
(5) A small circle is drawn at $(0.5, 0.4)$, and "World Peace" is written at the same place. This time, adj = c(1, 0) is set in text(). Here, the position set in text() is located in the lower right-hand corner of the text.
(6) A small circle is drawn at $(0.5, 0.2)$, and "World Peace" is written at the same place. In this case, adj = c(0, 1) is set in text(). The position specified in text() is placed in the lower left-hand corner of the text.

Text can be rotated as shown in Fig. 1.31 (right), which is produced by Program (1 - 15).

```
Program (1 - 15)
function() {
# (1)
  par(mai = c(1, 1, 1, 1), omi = c(0, 0, 0, 0), cex = 1.5)
# (2)
  plot(c(0, 1), c(0, 1), xlab = "", ylab = "", type = "n")
# (3)
  points(0.5, 0.5)
# (4)
  text(0.5, 0.5, "adj=-0.1", cex = 1, adj = -0.1, srt = 45)
  text(0.5, 0.5, "adj=-0.2", cex = 1, adj = -0.2, srt = 90)
  text(0.5, 0.5, "adj=-0.3", cex = 1, adj = -0.3, srt = 135)
  text(0.5, 0.5, "adj=-0.4", cex = 1, adj = -0.4, srt = 180)
  text(0.5, 0.5, "adj=-0.5", cex = 1, adj = -0.5, srt = 225)
  text(0.5, 0.5, "adj=-0.6", cex = 1, adj = -0.6, srt = 270)
```

```
  text(0.5, 0.5, "adj=-0.7", cex = 1, adj = -0.7, srt = 315)
  text(0.5, 0.5, "adj=-0.8", cex = 1, adj = -0.8, srt = 360)
}
```

(1) The graphics area is set.
(2) The coordinate axes are drawn.
(3) A small circle is drawn at $(0.5, 0.5)$.
(4) The settings of adj = and srt = in text() are modified in various ways to rotate the text.

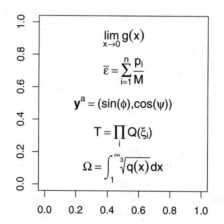

Figure 1.32   Equations written by Program (1 - 16).

The use of expression() enables mathematical equations to be written in the area set by plot() as shown in Fig. 1.32, which is constructed by Program (1 - 16)

```
Program (1 - 16)
function() {
# (1)
  par(mai = c(1, 1, 1, 1), omi = c(0, 0, 0, 0))
# (2)
  plot(c(0, 1), c(0, 1), xlab = "", ylab = "", type = "n")
# (3)
  text(0.5, 0.9, expression(lim(g(x), x %->% 0)), cex = 1.2)
# (4)
  text(0.5, 0.7, expression(bar(epsilon) == sum(frac(p[i], M),
    i==1, n)), cex = 1.2)
# (5)
  text(0.5, 0.5, expression(bold(y)^a == paste("(",plain(sin),
    "(", phi, ")",
    ",", plain(cos), "(", psi, "))" )),cex = 1.2)
```

```
# (6)
  text(0.5, 0.3, expression(T == paste(prod(plain(Q),i),
   "(", xi[i], ")")), cex = 1.2)
# (7)
  text(0.5, 0.1, expression(Omega == paste(integral(sqrt(q(x), 3),
   1, infinity), "dx")), cex = 1.2)
}
```

The execution of `help("plotmath")` in the console window of R shows the list of mathematical expressions that can be used as arguments of `expression()`.

1.8   VARIOUS POINTS AND STRAIGHT LINES

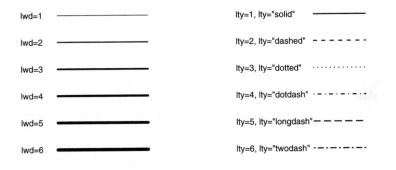

Figure 1.33   Straight lines with various widths drawn by Program (1 - 17) (left). Various types of straight lines drawn by Program (1 - 18) (right).

The R programs presented so far have not specified the widths of straight lines in `lines()`. Hence, the thinnest line is selected. To illustrate how the width of a straight line is specified, Program (1 - 17) is executed to draw Fig. 1.33 (left).

```
Program (1 - 17)
function() {
# (1)
  par(mai = c(1, 1, 1, 1), omi = c(0, 0, 0, 0))
# (2)
  plot(c(0, 1), c(0, 1), xlab = "", ylab = "", type = "n",
   axes = F)
# (3)
  text(0.1, 1, "lwd=1")
  lines(c(0.3, 1), c(1, 1), lwd = 1)
```

```
# (4)
   text(0.1, 0.8, "lwd=2")
   lines(c(0.3, 1), c(0.8, 0.8), lwd = 2)
# (5)
   text(0.1, 0.6, "lwd=3")
   lines(c(0.3, 1), c(0.6, 0.6), lwd = 3)
# (6)
   text(0.1, 0.4, "lwd=4")
   lines(c(0.3, 1), c(0.4, 0.4), lwd = 4)
# (7)
   text(0.1, 0.2, "lwd=5")
   lines(c(0.3, 1), c(0.2, 0.2), lwd = 5)
# (8)
   text(0.1, 0.0, "lwd=6")
   lines(c(0.3, 1), c(0.0, 0.0), lwd = 6)
}
```

(1) The graphics area is set.

(2) The coordinate axes are set but not labeled.

(3) "lwd=1" is written at $(0.1, 1)$. A straight line is drawn on the right side. lwd = 1 is specified as an argument of lines(). Even if lwd = is not set, the same straight line is obtained.

(4)(5)(6)(7)(8) Straight lines are drawn with various values of lwd = ranging from 2 to 6.

In addition, the type of straight line as well as its width can be specified as an argument of lines(). Program (1 - 18) illustrates this function as shown in Fig. 1.33 (right).

```
Program (1 - 18)
function() {
# (1)
   par(mai = c(1, 1, 1, 1), omi = c(0, 0, 0, 0))
# (2)
  plot(c(0,1), c(0,1), xlab = "", ylab = "", type = "n",
    axes = F)
# (3)
   text(0.0, 1, 'lty=1, lty="solid"', adj = c(0, 0.5))
   lines(c(0.6, 1), c(1, 1), lwd = 2, lty = "solid")
# (4)
   text(0.0, 0.8, 'lty=2, lty="dashed"', adj = c(0, 0.5))
   lines(c(0.6, 1), c(0.8, 0.8), lwd = 2, lty = "dashed")
# (5)
   text(0.0, 0.6, 'lty=3, lty="dotted"', adj = c(0, 0.5))
   lines(c(0.6, 1), c(0.6, 0.6), lwd = 2, lty = "dotted")
# (6)
   text(0.0, 0.4, 'lty=4, lty="dotdash"', adj = c(0, 0.5))
```

```
  lines(c(0.6, 1), c(0.4, 0.4), lwd = 2, lty = "dotdash")
# (7)
  text(0.0, 0.2, 'lty=5, lty="longdash"', adj = c(0, 0.5))
  lines(c(0.6, 1), c(0.2, 0.2), lwd = 2, lty = "longdash")
# (8)
  text(0.0, 0.0, 'lty=6, lty="twodash"', adj = c(0, 0.5))
  lines(c(0.6, 1), c(0.0, 0.0), lwd = 2, lty = "twodash")
}
```

(1) The graphics area is set.
(2) The coordinate axes are set but not labeled.
(3)(4)(5)(6)(7)(8) Straight lines are drawn with various values of `lty` = ranging from 1 to 6. These numbers in `lines()` can be replaced with words. The relationships between numbers and words are added.

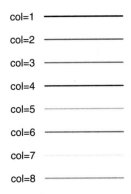

Figure 1.34   Straight lines of various colors given by Program (1 - 19).

Furthermore, the color of a straight line can be specified as an argument of `lines()` as demonstrated by Program (1 - 19), which constructs Fig. 1.34.

```
Program (1 - 19)
function() {
# (1)
  par(mai = c(1, 1, 1, 1), omi = c(0, 0, 0, 0))
# (2)
  plot(c(-0.4, 1), c(-0.4, 1), xlab = "", ylab = "", type = "n",
    axes = F)
# (3)
  text(0.1, 1, "col=1")
  lines(c(0.3, 1), c(1, 1), lwd = 2, col = 1)
# (4)
  text(0.1, 0.8, "col=2")
```

```
  lines(c(0.3, 1), c(0.8, 0.8), lwd = 2, col = 2)
# (5)
  text(0.1, 0.6, "col=3")
  lines(c(0.3, 1), c(0.6, 0.6), lwd = 2, col = 3)
# (6)
  text(0.1, 0.4, "col=4")
  lines(c(0.3, 1), c(0.4, 0.4), lwd = 2, col = 4)
# (7)
  text(0.1, 0.2, "col=5")
  lines(c(0.3, 1), c(0.2, 0.2), lwd = 2, col = 5)
# (8)
  text(0.1, 0.0, "col=6")
  lines(c(0.3, 1), c(0.0, 0.0), lwd= 2, col = 6)
# (9)
  text(0.1, -0.2, "col=7")
  lines(c(0.3, 1), c(-0.2, -0.2), lwd = 2, col = 7)
# (10)
  text(0.1, -0.4, "col=8")
  lines(c(0.3, 1), c(-0.4, -0.4), lwd = 2, col = 8)
}
```

(1) The graphics area is set.
(2) The coordinate axes are set but not labeled.
(3)(4)(5)(6)(7)(8)(9)(10) Straight lines are drawn with various values of col = ranging from 1 to 8.

The setting of colors in lines() can be done with words. The relationships between numbers and words are as follows.

Numbers	Words	Numbers	Words
col = 1	col = "black"	col = 5	col = "skyblue"
col = 2	col = "red"	col = 6	col = "magenta"
col = 3	col = "green"	col = 7	col = "yellow"
col = 4	col = "blue"	col = 8	col = "gray"

Grays of various densities can be used to color straight lines given by lines() as shown in Program (1 - 20), which outputs Fig. 1.35.

```
Program (1 - 20)
function() {
# (1)
  par(mai = c(1, 1, 1, 1), omi = c(0, 0, 0, 0))
# (2)
  plot(c(0, 1), c(0, 1), xlab = "", ylab = "", type = "n",
    axes = F)
# (3)
  text(0.0, 1, "col=gray(1)", adj = c(0, 0.5))
```

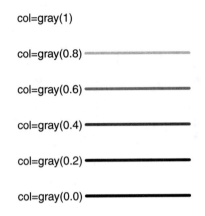

col=gray(1)

col=gray(0.8)

col=gray(0.6)

col=gray(0.4)

col=gray(0.2)

col=gray(0.0)

Figure 1.35   Gray straight lines of various densities, drawn by Program (1 - 20)

```
  lines(c(0.4, 1), c(1, 1), lwd = 4, col = gray(1))
# (4)
  text(0.0, 0.8, "col=gray(0.8)", adj = c(0, 0.5))
  lines(c(0.4, 1), c(0.8, 0.8), lwd = 4, col = gray(0.8))
# (5)
  text(0.0, 0.6, "col=gray(0.6)", adj = c(0, 0.5))
  lines(c(0.4, 1), c(0.6, 0.6), lwd = 4, col = gray(0.6))
# (6)
  text(0.0, 0.4, "col=gray(0.4)", adj = c(0, 0.5))
  lines(c(0.4, 1), c(0.4, 0.4), lwd = 4, col = gray(0.4))
# (7)
  text(0.0, 0.2, "col=gray(0.2)", adj = c(0, 0.5))
  lines(c(0.4, 1), c(0.2, 0.2), lwd = 4, col = gray(0.2))
# (8)
  text(0.0, 0.0, "col=gray(0.0)", adj = c(0, 0.5))
  lines(c(0.4, 1), c(0.0, 0.0), lwd = 4, col = gray(0.0))
}
```

(1) The graphics area is set.
(2) The coordinate axes are set but not labeled.
(3)(4)(5)(6)(7)(8)(9)(10) Straight lines are drawn with various values of `gray()` ranging from 0 to 1. The color of a straight line is white if `gray(1)` is set. Hence, a straight line is not drawn if the background color is white.

   `points()` not only draws small circles but also draws other symbols by specifying `pch =`. The R program below exemplifies this function and produces Fig. 1.36 (left).

Program (1 - 21)
```
function() {
```

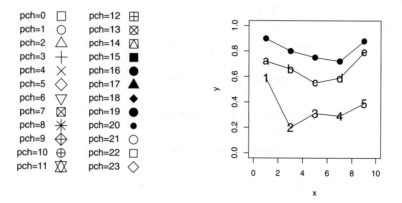

Figure 1.36   Various symbols given by Program (1 - 21) (left).   Line chart constructed by Program (1 - 22) (right).

```
# (1)
  par(mai = c(1, 1, 1, 1), omi = c(0, 0, 0, 0))
# (2)
  plot(c(0, 1), c(0, 1.1), xlab = "", ylab = "", type = "n",
  axes = F)
# (3)
  text(c(0.1, 0.1, 0.1, 0.1, 0.1, 0.1, 0.1, 0.1, 0.1, 0.1,
    0.1, 0.1), c(1.1, 1.0, 0.9, 0.8, 0.7, 0.6, 0.5, 0.4,
    0.3, 0.2, 0.1, 0), c("pch=0", "pch=1", "pch=2",
    "pch=3", "pch=4", "pch=5", "pch=6", "pch=7", "pch=8",
    "pch=9", "pch=10", "pch=11"), cex = 1)
# (4)
  points(c(0.3, 0.3, 0.3, 0.3, 0.3, 0.3, 0.3, 0.3, 0.3, 0.3,
    0.3, 0.3), c(1.1, 1.0, 0.9, 0.8, 0.7, 0.6, 0.5, 0.4, 0.3,
    0.2, 0.1, 0), pch=c(0, 1, 2, 3, 4, 5, 6, 7, 8, 9, 10, 11),
    cex = 2)
# (5)
 text(c(0.6, 0.6, 0.6, 0.6, 0.6, 0.6, 0.6, 0.6, 0.6, 0.6,
    0.6, 0.6), c(1.1, 1.0, 0.9, 0.8, 0.7, 0.6, 0.5, 0.4,
    0.3, 0.2, 0.1, 0), c("pch=12", "pch=13", "pch=14",
    "pch=15", "pch=16", "pch=17", "pch=18", "pch=19",
    "pch=20", "pch=21", "pch=22", "pch=23"), cex = 1)
# (6)
  points(c(0.8, 0.8, 0.8,0.8, 0.8, 0.8, 0.8, 0.8, 0.8, 0.8,
    0.8, 0.8), c(1.1, 1.0, 0.9, 0.8, 0.7, 0.6, 0.5, 0.4,
    0.3, 0.2, 0.1, 0), pch=c(12, 13, 14, 15, 16, 17, 18,
```

```
   19, 20, 21, 22, 23), cex = 2)
}
```

(1) The graphics area is set.
(2) The coordinate axes are set but not labeled.
(3)(4) Symbols given by various settings of **pch** = ranging from 0 to 11 are drawn alongside the settings.
(5)(6) Symbols given by various settings of **pch** = ranging from 12 to 23 are drawn alongside the settings.

A line chart such as Fig. 1.36 (right) can be illustrated. It is realized by combining the functions of **lines()**, which draws a straight line, and **points()**, which draws a symbol. The R program for this graph is Program (1 - 22).

Program (1 - 22)
```
function() {
# (1)
  par(mai = c(1, 1, 1, 1), omi = c(0, 0, 0, 0))
# (2)
  plot(c(0, 10), c(0, 1), xlab = "x", ylab = "y", type = "n")
# (3)
  xx <- c(1, 3, 5, 7, 9)
  yy1 <- c(0.9, 0.8, 0.75, 0.72, 0.88)
  yy2 <- c(0.72, 0.66, 0.55, 0.59, 0.79)
  yy3 <- c(0.59, 0.2, 0.31, 0.29, 0.39)
# (4)
  points(xx, yy1, pch = 16, cex = 1.5)
  lines(xx, yy1)
# (5)
  points(xx, yy2, pch = letters[1:5], cex = 1.5)
  lines(xx, yy2)
# (6)
  points(xx, yy3, pch = c("1", "2", "3", "4", "5"), cex = 1.5)
  lines(xx, yy3)
}
```

(1) The graphics area is set.
(2) The coordinate axes are set, drawn, and labeled.
(3) Values are stored in variables (objects) specifying the positions on a graph. **xx** stores values for positions on the x-axis. **yy1**, **yy2**, and **yy3** store values for positions on the y-axis.
(4) Straight lines and small filled circles based on **xx** and **yy1** are drawn.
(5) Straight lines and letters based on **xx** and **yy2** are drawn. **letters[1:5]** functions in the same way as **c("a", "b", "c", "d", "e")**. The text specified by **pch** = must comprise one letter.
(6) Specific letters are set for **pch** = as an argument of **points()**.

1.9   FONTS

Various fonts are available for displaying text in graphs. The type of font available depends on the setting of fonts in the OS installed on a PC. Additionally, the fonts shown on a display may be different from those in a digital file saved in postscript format. That is, the fonts shown on a display cannot be stored as they are in a digital file on some occasions.

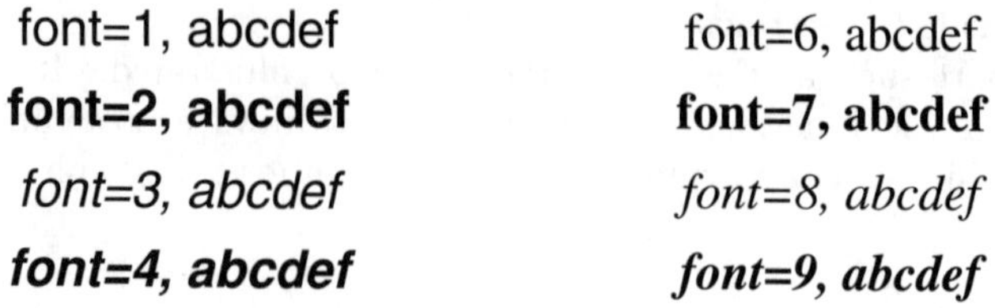

font=1, abcdef   font=6, abcdef

font=2, abcdef   **font=7, abcdef**

font=3, abcdef   *font=8, abcdef*

font=4, abcdef   ***font=9, abcdef***

Figure 1.37   Text in Helvetica plain, Helvetica bold, Helvetica italic, and Helvetica bold italic in descending order, given by Program (1 - 23) (left). Text in Times plain, Times bold, Times italic, and Times bold italic in descending order, given by Program (1 - 23) (right).

Fig. 1.37 (left) employs Helvetica fonts. Texts in Helvetica plain, Helvetica bold, Helvetica italic, and Helvetica bold italic in descending order, constructed by Program (1 - 23), are illustrated.

Program (1 - 23)

```
function() {
# (1)
  par(mai = c(1, 1, 1, 1), omi = c(0, 0, 0, 0))
# (2)
  plot(c(0, 1), c(0, 1), xlab = "", ylab = "", type = "n",
  axes = F)
# (3)
  text(0.5, 0.8, "font=1, abcdef", cex = 2, font = 1)
# (4)
  text(0.5, 0.6, "font=2, abcdef", cex = 2, font = 2)
# (5)
  text(0.5, 0.4, "font=3, abcdef", cex = 2, font = 3)
# (6)
  text(0.5, 0.2, "font=4, abcdef", cex = 2, font = 4)
}
```

(1) The graphics area is set.
(2) The coordinate axes are set.
(3) The command text() draws "font=1, abcdef". The argument font = 1 indicates Helvetica plain. However, the meaning of font = 1 depends on the setting of fonts in the OS on a PC.

(4) "font=2, abcdef" is written. The argument `font = 2` indicates Helvetica bold.

(5) "font=3, abcdef" is written. The argument `font = 3` indicates Helvetica italic.

(6) "font=4, abcdef" is written. The argument `font = 4` indicates Helvetica bold italic.

Fig. 1.37 (right), constructed by Program (1 - 24), employs the Times font.

```
Program (1 - 24)
function() {
# (1)
  par(mai = c(1, 1, 1, 1), omi = c(0, 0, 0, 0))
# (2)
  plot(c(0, 1), c(0, 1), xlab = "", ylab = "", type = "n",
    axes = F)
# (3)
  text(0.5, 0.8, "font=6, abcdef", cex = 2, font = 6)
# (4)
  text(0.5, 0.6, "font=7, abcdef", cex = 2, font = 7)
# (5)
  text(0.5, 0.4, "font=8, abcdef", cex = 2, font = 8)
# (6)
  text(0.5, 0.2, "font=9, abcdef", cex = 2, font = 9)
}
```

(1) The graphics area is set.

(2) The coordinate axes are set.

(3) "font=6, abcdef" is written. The argument `font = 6` indicates Times plain. The meaning of `font = 1` depends on the setting of the OS installed on a PC.

(4) "font=7, abcdef" is written. The argument `font = 7` indicates Times bold.

(5) "font=8, abcdef" is written. The argument `font = 8` indicates Times italic.

(6) "font=9, abcdef" is written. The argument `font = 9` indicates Times bold italic.

1.10   FIGURES SUCH AS CIRCLES AND RECTANGLES

The R command `symbols()` draws circles, rectangles, stars, thermometers, and boxplots. Fig. 1.38 (left), constructed by Program (1 - 25), exemplifies figures given by `symbols()`.

```
Program (1 - 25)
function() {
```

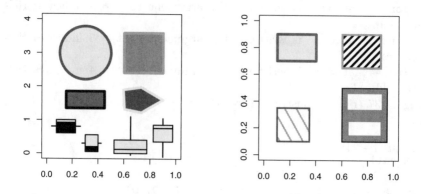

Figure 1.38   Various figures given by Program (1 - 25) (left). Rectangles given by Program (1 - 26) (right).

```
# (1)
  par(mai = c(1, 1, 1, 1), omi = c(0, 0, 0, 0))
# (2)
  plot(c(0, 1), c(0, 4), xlab = "", ylab = "", type = "n")
# (3)
  symbols(0.3, 3, add = T, inches = F, circles = 0.2,
    fg = "red", bg = "yellow", lwd = 5)
# (4)
  symbols(0.75, 3, add = T, inches = F, squares = 0.3,
    fg = "green", bg = "skyblue", lwd = 5)
# (5)
  symbols(0.3, 1.6, add = T, inches = F, rectangles =
    matrix(c(0.3, 0.5), nrow = 1), fg = "blue", bg = "magenta",
    lwd = 5)
# (6)
  symbols(0.7, 1.6, add = T, inches = F, stars = matrix(
    c(0.2, 0.1, 0.13, 0.12, 0.1), nrow = 1), fg = "yellow",
    bg = "red", lwd = 5)
# (7)
  th1 <- matrix(c(0.15, 0.1, 0.4, 0.5, 0.8, 0.3), ncol = 3)
  symbols(c(0.15, 0.35), c(0.8, 0.3), add = T, inches = F,
    thermometers = th1, fg = "blue", bg = "yellow", lwd = 2)
# (8)
  box1 <- matrix(c(0.25, 0.15, 0.4, 0.5, 0.1, 0.4, 0.1, 0.2,
    0.3, 0.8), ncol = 5)
  symbols(c(0.65, 0.9),c(0.2, 0.6), add = T, inches = F,
    boxplots = box1, fg = "blue", bg = "yellow", lwd = 2)
```

}

(1) The graphics area is set.

(2) The coordinate axes are set.

(3) Circles are drawn using `symbols()`. `add = T` indicates adding a figure to an existing graph (the coordinate axes in this case). The setting of `add = F` or the default setting means drawing a new graph. The argument `inches = F` specifies the method of setting the size of a circle. The argument `inches = F` means that the radius of a circle is specified on the scale of the x-axis. The argument `circles = 0.2` means the radius of a circle. The argument `fg = "red"` indicates that the circumference of a circle is drawn in red. The argument `bg = "yellow"` shows that the inner area of a circle is drawn in yellow.

(4) The command `symbols()` draws a square. The argument `inches = F` specifies the method of setting the size of a square. The argument `inches = F` means that the length of the side of a square is specified on the scale of x-axis. The argument `squares = 0.3` sets the length of the side of a square.

(5) The command `symbols()` draws a rectangle. The argument `inches = F` specifies the method of setting the length of the sides of a rectangle. The argument `inches = F` means that the length of the sides of a rectangle are specified on the scales of the x-axis and y-axis. The argument `rectangles = matrix(c(0.3, 0.5), nrow = 1)` indicates that the horizontal width of the rectangle is 0.3 (on the scale of x-axis), and the vertical length is 0.5 (on the scale of the y-axis).

(6) The command `symbols()` draws a star. A star is given by specifying the position of the original point, dividing 360 degrees equally into a specific number of directions on the basis of the original point, drawing radial straight lines of specific lengths along these directions, and connecting the end points of the lines by straight lines. The argument `inches = F` specifies the method of setting the lengths of radial straight lines constituting a star. The first element indicates the length of the rightward straight line from the original point. Other elements specify the lengths of the straight lines in a counterclockwise direction. The argument `inches = F` indicates that the lengths of radial straight lines constituting a star are specified on the scale of the x-axis. Since `stars = matrix(c(0.2, 0.1, 0.13, 0.12, 0.1), nrow = 1)` gives a matrix with five elements, it indicates that the shape of a star with five apexes is illustrated; each element of the matrix specifies the length of the straight line in each direction.

(7) The command `symbols()` draws a thermometer. The argument `inches = F` specifies the method of setting the lengths of the sides of a thermometer and its shape. The argument `inches = F` indicates that the lengths of the sides of a thermometer are set on the scales of the x-axis and y-axis. The argument `thermometers = th1` specifies the lengths of the sides of a thermometer and its shape. The elements of `th1` in this example are as follows:

    [,1] [,2] [,3]

```
[1,] 0.15  0.4  0.8
[2,] 0.10  0.5  0.3
```

The element of 0.15 is the horizontal length of the left thermometer on the scale of the x-axis. The element of 0.4 is the vertical length of the left thermometer on the scale of the y-axis. The element of 0.8 shows the ratio of the lower part of the left thermometer. The element of 0.10 is the horizontal length of the right thermometer on the scale of the x-axis. The element of 0.5 is the vertical length of the right thermometer on the scale of the y-axis. The element of 0.3 indicates the ratio of the lower part of the right thermometer. (8) The command symbols() draws a boxplot. The argument inches = F indicates that the lengths of the sides and whiskers of a boxplot is specified on the scales of the x-axis and y-axis. The argument boxplots = box1 specifies the lengths of the sides and whiskers of a boxplot. box1 in this example is as follows:

```
     [,1]  [,2]  [,3]  [,4]  [,5]
[1,] 0.25  0.4   0.05  0.7   0.3
[2,] 0.15  0.5   0.45  0.2   0.8
```

The element of 0.25 is the horizontal width of the left boxplot on the scale of the x-axis. The element of 0.4 is the vertical width of the left boxplot on the scale of the y-axis. The element of 0.05 is the length of the lower whisker of the left boxplot. The element of 0.7 is the length of the upper whisker of the left boxplot. The element of 0.3 indicates the position of the middle line of the left boxplot: the ratio of the lower part of the box under the middle line. The element of 0.15 is the horizontal width of the right boxplot on the scale of the x-axis. The element of 0.5 is the vertical width of the right boxplot on the scale of the y-axis. The element of 0.45 is the length of the lower whisker of the right boxplot. The element of 0.2 is the length of the upper whisker of the right boxplot. The element of 0.8 indicates the position of the middle line of the right boxplot: the ratio of the lower part of the box under the middle line.

The command rect() allows drawing of various rectangles. For example, Program (1 - 26) gives Fig. 1.38 (right).

```
Program (1 - 26)
function(){
# (1)
  par(mai = c(1, 1, 1, 1), omi = c(0, 0, 0, 0))
# (2)
  plot(c(0, 1), c(0, 1), xlab = "", ylab = "", type = "n")
# (3)
  rect(0.1, 0.7, 0.4, 0.9, border = "magenta", col = "yellow",
    lwd = 5)
# (4)
```

```
  rect(0.6, 0.65, 0.9, 0.9, border = "green", col = "blue",
    density = 10, ljoin = 1, lwd = 4)
# (5)
  rect(0.1, 0.1, 0.35, 0.35, border = "red", col = "green",
    density = 5, angle = 120, ljoin = 1, lwd = 3)
# (6)
  rect(0.6, 0.1, 0.95, 0.5, border= "blue", col = "skyblue",
    ljoin = 1,   lwd = 2)
# (7)
  rect(0.65, 0.35, 0.9, 0.45, border = "white", col = "white",
    ljoin = 1, lwd = 4)
# (8)
  rect(0.65, 0.15, 0.9, 0.25, border = "transparent",
    col = "white", ljoin = 1, lwd = 4)
}
```

(1) The graphics area is set.
(2) The coordinate axes are drawn.
(3) The command rect() draws a rectangle. The argument border =
"magenta" indicates that the rectangle is bordered in magenta. Since ljoin =
is not specified, ljoin = 0 is assumed. If ljoin = 0 is set, the contact points
between two straight lines are round. The setting of ljoin = "round" gives
the same result.
(4) The command rect() draws a rectangle. If ljoin = 1 is set, the contact
points between two straight lines are acute. The setting of ljoin = "mitre"
gives the same result. Since density = 10 is set, the rectangular region is
striped with 10 diagonal lines per inch. Since angle = is not set, angle = 45
is assumed. Therefore, the inclination of the diagonal lines is 45 degrees
when the angle is measured in a counterclockwise direction from the rightward
straight line drawn from the original point.
(5) The command rect() draws a rectangle. Since density = 5 is set,
the rectangular region is striped with five diagonal lines per inch. Since
angle = 120 is set, the inclination of the diagonal lines is 120 degrees when
the angle is measured in a counterclockwise direction from the straight line
drawn rightward from the original point.
(6)(7)(8) The command rect() draws a skyblue rectangle and two white
rectangles in it. The sizes of the two white rectangles are set to be the same.
However, since the upper rectangle is given by border = "white" and the
lower rectangle is given by border = "transparent", the upper rectangle is
slightly larger than the lower one.
    polygon() draws polygons in a concise manner. For instance, Program
(1 - 27) draws Fig. 1.39.

Program (1 - 27)
```
function() {
# (1)
```

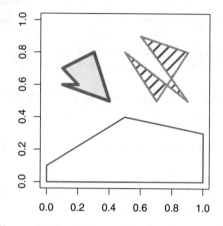

Figure 1.39   Polygon given by Program (1 - 27).

```
  par(mai = c(1, 1, 1, 1), omi = c(0, 0, 0, 0))
# (2)
  plot(c(0, 1), c(0, 1), xlab = "", ylab = "", type = "n")
# (3)
  polygon(c(0.1, 0.4, 0.3, 0.1, 0.2), c(0.6, 0.5, 0.8, 0.7,
    0.6), border = "magenta", col = "yellow", lwd = 5)
# (4)
  polygon(c(0.6, 0.9, 0.5, 0.7, 0.9), c(0.9, 0.5, 0.8, 0.5, 0.8),
    border = "skyblue", col = "red", density = 8, lwd = 3)
# (5)
  polygon(c(0.0, 0.5, 1, 1, 0.5, 0), c(0.1, 0.4, 0.3, 0, 0, 0),
    density = 0, col = "red", lwd = 2)
}
```

(1) The graphics area is set.

(2) The coordinate axes are drawn.

(3) The command `polygon()` draws a polygon. The given points are connected sequentially and a closed polygon is formed by connecting the first point and last point.

(4) The command `polygon()` draws a polygon. When lines connecting the given points cross, a figure the same as that in Fig. 1.39 is formed.

(5) The command `polygon()` draws a polygon. Since `density = 0` is set, the inside of the rectangle is neither filled nor striped.

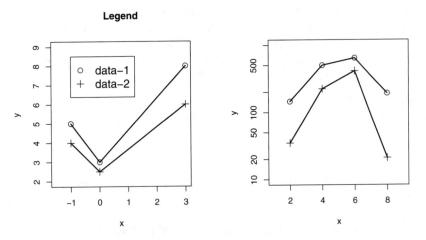

Figure 1.40 Graph with a legend, given by Program (1 - 28) (left). Line chart given by Program (1 - 29) (right).

1.11 LEGENDS AND LOGARITHMIC PLOTS

A legend can be added in a graph. For example, the use of legend() in the following program yields Fig. 1.40 (left).

Program (1 - 28)
```
function() {
# (1)
  par(mai = c(1, 1, 1, 1), omi = c(0, 0, 0, 0))
# (2)
  plot(c(-1.5, 3), c(2, 9), xlab = "x", ylab = "y", type = "n",
  main = "Legend")
# (3)
  lines(c(-1, 0, 3), c(5, 3, 8), lwd = 2)
  points(c(-1, 0, 3), c(5, 3, 8), pch = 1, cex =1.3)
  lines(c(-1, 0, 3), c(4, 2.5, 6), lwd = 2)
  points(c(-1, 0, 3), c(4, 2.5, 6), pch = 3, cex = 1.3)
# (4)
  legend(-1, 8.5, legend = c("data-1", "data-2"), pch = c(1, 3),
  cex = 1.3)
}
```

(1) The graphics area is set.
(2) The coordinate axes are drawn. The argument main = "Legend" specifies the title of the graph.
(3) Straight lines, circles, and x-indications are drawn.

(4) A legend is put at $(-1, 8.5)$. The elements of legend("data-1", "data-2") give the text for respective lines. The elements of pch = c(1, 3) specify the symbols corresponding to respective text in the legend.

The logarithmic scale can be used as the scale of the axes in a graph. For example, Program (1 - 29) depicts Fig. 1.40 (right).

```
Program (1 - 29)
function() {
# (1)
  par(mai = c(1, 1, 1, 1), omi = c(0, 0, 0, 0))
# (2)
  plot(c(1, 9), c(10, 1000), xlab = "x", ylab = "y", type = "n",
  log = "y")
# (3)
  lines(c(2, 4, 6, 8), c(145, 500, 641, 192), lwd = 2)
  points(c(2, 4, 6, 8), c(145, 500, 641, 192), pch = 1,
  cex = 1.3)
# (4)
  lines(c(2, 4, 6, 8), c(35, 223, 412, 21), lwd = 2)
  points(c(2, 4, 6, 8), c(35, 223, 412, 21), pch = 3, cex = 1.3)
}
```

(1) The graphics area is set.
(2) The coordinate axes are drawn. Since log = "y" is set in plot(), the scale of the y-axis is the logarithmic scale. The argument log = "x" makes the scale of the x-axis the logarithmic scale. The argument log = "xy" makes both scales the logarithmic scales.
(3) Straight lines and circles are drawn.
(4) Straight lines and x-indications are drawn.

1.12   BAR CHARTS

The command barplot() illustrates a bar plot. For example, Program (1 - 30) gives Fig. 1.41 (left).

```
Program (1 - 30)
function() {
# (1)
  par(mai = c(1, 1, 1, 1), omi = c(0, 0, 0, 0))
# (2)
  yy <- c(225, 500, 641, 192)
# (3)
  name1 <- c("data-a", "data-b", "data-c", "data-d")
# (4)
  barplot(yy, ylab = "y", names.arg = name1)
```

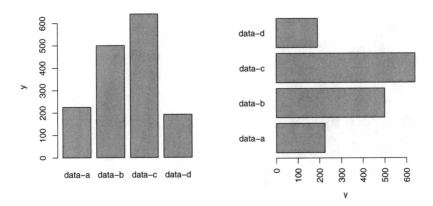

Figure 1.41   Horizontal bar plot given by Program (1 - 30) (left). Horizontal bar plot given by Program (1 - 31) (right).

```
}
```

(1) The graphics area is set.
(2) Data for four bars are stored as yy.
(3) The names of the four bars are set in name1.
(4) The command barplot() illustrates a bar plot for representing the values of yy. The argument names.arg = name1 specifies the names of the four bars.

   A bar plot illustrated by barplot() can be horizontal. For example, if (4) of Program (1 - 30) is replaced with the following command, Fig. 1.41 (left) is obtained.

```
# (4)
  barplot(yy,  xlab = "y", names.arg = name1, horiz = T, las = 2)
```

The command barplot() draws a bar plot for representing the values of yy. The argument horiz = T makes the graph horizontal. The argument las = 2 makes the names of the respective bars horizontal.

   The bar plot given by barplot() can be a bar plot with partitioned bars. For example, Program (1 - 31) realizes Fig. 1.42 (left).
Program (1 - 31)

```
function() {
# (1)
  par(mai = c(1, 1, 1, 1), omi = c(0, 0, 0, 0))
# (2)
  yy1 <- c(145, 239, 640, 192)
  yy2 <- c(321, 233, 204, 218)
# (3)
```

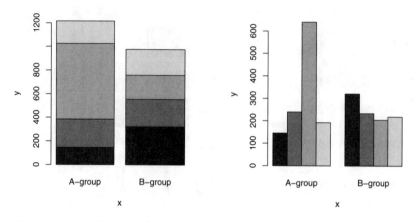

Figure 1.42   Partitioned bar plot given by Program (1 - 31) (left). Grouped bar plot given by Program (1 - 32) (right).

```
  yy <- cbind(yy1, yy2)
# (4)
  barplot(yy, xlab = "x", ylab = "y", names.arg = c("A-group",
  "B-group"))
}
```

(1) The graphics area is set.

(2) The values of the respective partitions of the first bar are given as the elements of **yy1**. The values of the respective partitions of the second bar are given as the elements of **yy2**.

(3) The matrix **yy** is constructed; the first column is **yy1** and the second column is **yy2**. The elements of **yy** here are as follows:

```
     yy1 yy2
[1,] 145 321
[2,] 239 233
[3,] 640 204
[4,] 192 218
```

(4) The command **barplot()** illustrates a bar plot for representing the values of **yy**. The argument **names.arg** =c("A-group", "B-group") specifies the names of respective bars.

   If the data of a bar plot given by **barplot()** is divided into groups, a grouped bar plot is a good option. For example, by substituting (4) of Program (1 - 31) with the following command, Fig. 1.42 (right) is obtained.

Program (1 - 32)

```
# (4)
  barplot(yy, xlab = "x", ylab = "y", beside = T, names.arg =
  c("A-group", "B-group"))
```

The program `barplot()` illustrates a bar plot for depicting the values of yy. The argument `names.arg = c("A-group", "B-group")` sets the names of the respective bars. The argument `beside = T` specifies that the two bar plots are positioned horizontally.

1.13 PIE CHARTS

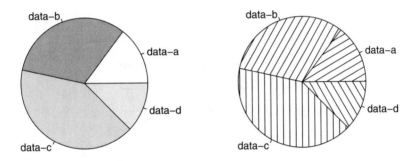

Figure 1.43   Pie chart given by Program (1 - 33) (left). Pie chart given by Program (1 - 34) (right).

A pie chart is drawn by a R program. For example, Program (1 - 33) using `pie()` yields Fig. 1.43 (left).

```
Program (1 - 33)
function() {
# (1)
  par(mai = c(1, 1, 1, 1), omi = c(0, 0, 0, 0))
# (2)
  yy <- c(225, 500, 641, 192)
# (3)
  name1 <- c("data-a", "data-b", "data-c", "data-d")
# (4)
  pie(yy, labels = name1)
}
```

(1) The graphics area is set.
(2) Data is stored as yy.
(3) The names of data is set as **name1**.
(4) The command `pie()` realizes a pie chart using yy.

The respective areas in Fig. 1.43 (right) are striped. This pie chart is realized by replacing (4) of Program (1 - 33) with the following command.

Program (1 - 34)
```
# (4)
  pie(yy, labels = name1, density = 10, angle = c(30, 60, 90, 120))
```

The argument `density =` in `pie()` sets the densities of the stripes. The argument `angle =` designates the angles of respective areas.

1.14   LAYOUT OF MULTIPLE GRAPHS

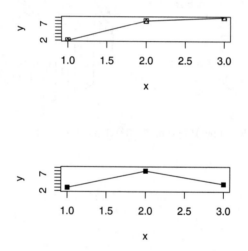

Figure 1.44   Use of `par(mfrow = c(2, 1), mai = c(1, 1, 1, 1), omi = c(0, 0, 0, 0))` (Program (1 - 35)).

Plural graphs can be illustrated in a graphics window. For example, Program (1 - 35) constructs Fig. 1.44.

Program (1 - 35)
```
function() {
# (1)
  par(mfrow = c(2, 1), mai = c(1, 1, 1, 1),
    omi = c(0, 0, 0, 0))
# (2)
```

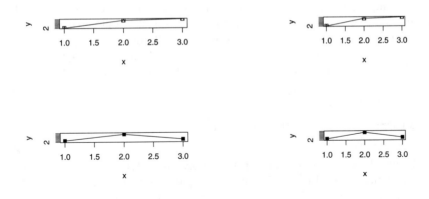

Figure 1.45 Use of `par(mfrow = c(2, 1)`, `mai = c(1.5, 1, 1, 1)`, `omi = c(0, 0, 0, 0))` (left). Use of `par(mfrow = c(2, 1)`, `mai = c(1.5, 2, 1, 1)`, `omi = c(0, 0, 0, 0))` (right).

```
  plot(c(1, 3), c(2, 9), xlab = "x", ylab = "y", type = "n")
  lines(c(1,2,3), c(2,8,9))
  points(c(1,2,3), c(2,8,9), pch = 14)
# (3)
  plot(c(1, 3), c(2, 9), xlab = "x", ylab = "y", type = "n")
  lines(c(1, 2, 3), c(3, 8, 4))
  points(c(1, 2, 3), c(3, 8, 4), pch = 15)
}
```

(1) The argument `mfrow = c(2, 1)` in `par()` specifies that the graphics window is divided into two equal-spaced areas vertically for drawing a graph in each area. The graphics area for each area is set.
(2) The first graph is drawn.
(3) The second graph is drawn.

When graphs are drawn in divided areas, the implications of `mai =` and `omi =` in `par()` are clear. To understand the indication of the first element of `mai =`, (1) of Program (1 - 35) is replaced with the command below; Fig. 1.45 (left) is obtained.

Program (1 - 36)

```
# (1)
  par(mfrow = c(2, 1), mai = c(1.5, 1, 1, 1),
    omi = c(0, 0, 0, 0))
```

The argument mai = c(1.5, 1, 1, 1) is set in par(). Hence, when Fig. 1.44 is compared with Fig. 1.45 (left), the figure margins for describing the explanation of the x-axis are wider in the two graphs located vertically in Fig. 1.45 (left). Therefore, the figure areas are vertically narrower.

To understand the second element of mai =, (1) of Program (1 - 36) is replaced with the following command; Fig. 1.45 (right) is obtained.

Program (1 - 37)
```
# (1)
  par(mfrow = c(2, 1), mai = c(1.5, 2, 1, 1),
    omi = c(0, 0, 0, 0))
```

The graphics area is set. The argument mai = c(1.5, 2, 1, 1) is set in par(). Hence, when Fig. 1.45 (left) is compared with Fig. 1.45 (right), the figure margins for describing the explanation of the y-axis are wider in the two graphs located vertically in Fig. 1.45 (right). Therefore, the figure areas are horizontally narrower.

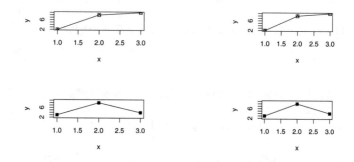

Figure 1.46   Use of par(mfrow = c(2, 1), mai = c(1.5, 2, 0.5, 0.5), omi = c(0.7, 0, 0, 0)) (left). Use of omi = c(0.7, 0, 0, 0)) mai = c(1.5, 2, 0.5, 0.5), omi = c(0.7, 0.7, 0, 0)) (right).

Furthermore, if the setting of omi = in par() is altered, the implications of omi = are clear. To understand the first element of omi =, (1) of Program (1 - 37) is replaced with the command below; Fig. 1.46 (left) is obtained.

Program (1 - 38)
```
# (1)
  par(mfrow = c(2, 1), mai = c(1.5, 2, 0.5, 0.5),
    omi = c(0.7, 0, 0, 0))
```

The argument omi = c(0.7, 0, 0, 0) is set in par(). Hence, the outer margin of the bottom area is wide in Fig. 1.46 (left). Therefore, the area for the two graphs is vertically narrower.

To understand the second element of `omi =`, (1) of Program (1 - 38) is replaced with the command below; Fig. 1.46 (right) is obtained.

Program (1 - 39)
```
# (1)
  par(mfrow = c(2, 1), mai = c(1.5, 2, 0.5, 0.5),
    omi = c(0.7, 0.7, 0, 0))
```

The argument `omi = c(0.7, 0.7, 0, 0)` is set in `par()`. Hence, when Fig. 1.46 (left) is compared with Fig. 1.46 (right), the outer margin of the left area is wider in Fig. 1.46 (right). Therefore, the area for the two graphs is horizontally narrower.

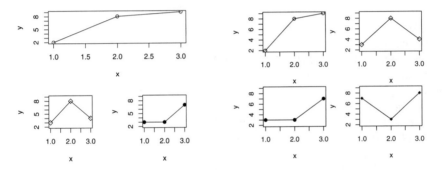

Figure 1.47   Three graphs given by Program (1 - 40) (left). Four graphs given by Program (1 - 41) (right).

After a graphics window is divided vertically, the bottom area can be divided horizontally. For example, Program (1 - 40) yields Fig. 1.47 (left).

Program (1 - 40)
```
function() {
# (1)
  fun1 <- function(){
    plot(c(1, 3), c(2, 9), xlab = "x", ylab = "y", type = "n")
    lines(c(1, 2, 3), c(2, 8 ,9))
    points(c(1, 2, 3), c(2, 8, 9), pch = 1)
  }
# (2)
  fun2 <- function(){
    plot(c(1, 3), c(2, 9), xlab = "x", ylab = "y", type= "n")
    lines(c(1, 2, 3), c(3, 8, 4))
```

```
      points(c(1, 2, 3), c(3, 8, 4), pch = 5)
   }
# (3)
   fun3 <- function(){
      plot(c(1, 3), c(2, 9), xlab = "x", ylab = "y", type = "n")
      lines(c(1, 2, 3), c(3, 3, 7))
      points(c(1, 2, 3), c(3, 3, 7), pch = 16)
   }
# (4)
   par(mfrow = c(1, 1), omi = c(0, 0, 0, 0))
# (5)
   split.screen(figs = c(2, 1))
   split.screen(figs = c(1, 2), screen = 2)
# (6)
   screen(1)
   par(mai = c(1.0, 1.0, 0.2, 0.2))
   fun1()
# (7)
   screen(3)
   par(mai = c(1.0, 1.0, 0.2, 0.2))
   fun2()
# (8)
   screen(4)
   par(mai = c(1.0, 1.0, 0.2, 0.2))
   fun3()
}
```

(1) The function fun1() for drawing the first graph is defined.

(2) The function fun2() for drawing the second graph is defined.

(3) The function fun3() for drawing the third graph is defined.

(4) The graphics area is set. Although the argument omi = is set in par(), mai = is not set. This is because mai = is specified for the three graphs.

(5) The command split.screen(figs = c(2, 1)) divides the graphics area vertically. The command split.screen(figs = c(1, 2), screen = 2) divides the second part of the dual-partitioned window horizontally. Although figs = is set in split.screen(figs = c(2, 1)), screen = is not set. Because the graphics window is not separated at this stage, screen = 0 is set automatically.

(6) The command screen(1) describes drawing the first graph. The argument mai = specifies the figure margin in the first graph. Then, the first graph is drawn.

(7) The command screen(3) describes drawing the second graph. It should be noted that this is not screen(2). The argument mai = specifies the figure margin in the second graph. Then, the second graph is drawn.

(8) The command `screen(4)` describes drawing the third graph. It should be noted that this is `screen(4)`, not `screen(3)`. The argument `mai` = specifies the figure margin in the third graph. Then, the third graph is drawn.

Four graphs can be placed on a grid. After the graphics window is divided into two equal-spaced areas vertically, each space is divided into two equal-spaced areas horizontally. For example, Program (1 - 41) realizes Fig. 1.47 (right).

Program (1 - 41)

```
function() {
# (1)
  fun1 <- function(){
    plot(c(1, 3), c(2, 9), xlab = "x", ylab = "y", type = "n")
    lines(c(1, 2, 3), c(2, 8, 9))
    points(c(1, 2, 3), c(2, 8, 9), pch = 1)
  }
# (2)
  fun2 <- function(){
    plot(c(1, 3), c(2, 9), xlab = "x", ylab = "y", type = "n")
    lines(c(1, 2, 3), c(3, 8, 4))
    points(c(1, 2, 3), c(3, 8, 4), pch = 5)
  }
# (3)
  fun3 <- function(){
    plot(c(1, 3), c(2, 9), xlab = "x", ylab = "y", type = "n")
    lines(c(1, 2, 3), c(3, 3, 7))
    points(c(1, 2, 3), c(3, 3, 7), pch = 16)
  }
# (4)
  fun4 <- function(){
    plot(c(1, 3), c(2, 9), xlab = "x", ylab = "y", type = "n")
    lines(c(1, 2, 3), c(7, 3, 8))
    points(c(1, 2, 3), c(7, 3, 8), pch = 18)
  }
# (5)
  par(mfrow = c(2, 2), mai = c(1, 1, 0, 0),
    omi = c(0.5, 0.5, 0.5, 0.5))
# (6)
  fun1()
# (7)
  fun2()
# (8)
  fun3()
# (9)
  fun4()
```

```
}
```

(1) The function **fun1()** for drawing the first graph is defined.
(2) The function **fun2()** for drawing the second graph is defined.
(3) The function **fun3()** for drawing the third graph is defined.
(4) The function **fun4()** for drawing the fourth graph is defined.
(5) The graphics area is set. Since the argument **mfrow =c(2, 2)** is set in **par()**, the whole area of the graphics window is divided into two equal-spaced areas vertically and each space is divided into two equal-spaced areas horizontally.
(6) The first graph is drawn.
(7) The second graph is drawn.
(8) The third graph is drawn.
(9) The fourth graph is drawn.

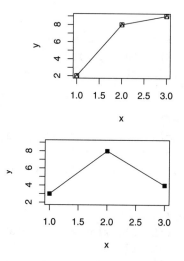

Figure 1.48   Two graphs given by Program (1 - 42).

The graphics windows can be divided into unequal-spaced areas as well as into equal-spaced areas. Program (1 - 42), which yields Fig. 1.48, exemplifies it.

```
Program (1 - 42)
function() {
# (1)
  fun1 <- function(){
    plot(c(1, 3), c(2, 9), xlab = "x", ylab = "y", type = "n")
    lines(c(1, 2, 3), c(2, 8, 9))
```

```
    points(c(1, 2, 3), c(2, 8, 9), pch = 14)
  }
# (2)
  fun2 <- function(){
    plot(c(1, 3), c(2, 9), xlab = "x", ylab = "y", type = "n")
    lines(c(1, 2, 3), c(3, 8, 4))
    points(c(1, 2, 3), c(3, 8, 4), pch = 15)
  }
# (3)
  par(mfrow = c(1, 1), omi = c(0, 0, 0, 0))
# (4)
  figs1 <-  matrix(c(0.1, 0.9, 0.6, 1.0, 0.1, 0.9, 0.0, 0.7),
    nrow = 2, byrow = T)
  print(figs1)
  split.screen(figs = figs1, erase = T)
# (5)
  screen(1)
  par(mai = c(1, 1.2, 0.1, 0.1))
  fun1()
# (6)
  screen(2)
  par(mai = c(1, 0.7, 0.1, 0.1))
  fun2()
}
```

(1) The function **fun1**() for drawing the first graph is defined.
(2) The function **fun2**() for drawing the second graph is defined.
(3) The graphics area is set. The argument **omi** = is set in **par**(). However,
mai = is not set.
(4) The matrix **figs1** for specifying the ratios of the division areas is defined.
The matrix **figs1** in this example is:

```
      [,1] [,2] [,3] [,4]
[1,]   0.1  0.9  0.6  1.0
[2,]   0.1  0.9  0.0  0.7
```

The first row represents the area of the first graph. The second row represents
the area of the second graph. The first two elements of each row stand for the
range of the x-axis. The last two elements of each row stand for the range of
the y-axis. Hence, two graphs positioned vertically are obtained. The whole
ranges of both the x-axis and the y-axis are between 0 and 1
(5) The first graph is drawn.
(6) The second graph is drawn.

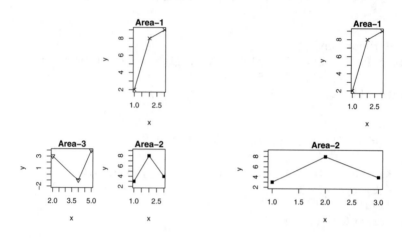

Figure 1.49   Three graphs given by Program (1 - 43) (left). Two graphs given by Program (1 - 44) (right).

When the graphics area is divided into two areas horizontally and vertically to locate the four graphs on a grid, the ratio of the division can be flexible. Fig. 1.49 (left) given by Program (1 - 43) is obtained.

Program (1 - 43)

```
function() {
# (1)
  fun1 <- function(){
    plot(c(1, 3), c(2, 9), xlab = "x", ylab = "y", type = "n",
    main = "Area-1")
    lines(c(1, 2, 3), c(2, 8, 9))
    points(c(1, 2, 3), c(2, 8, 9), pch = 4)
  }
# (2)
  fun2 <- function(){
    plot(c(1, 3), c(2, 9), xlab = "x", ylab = "y", type = "n",
    main = "Area-2")
    lines(c(1, 2, 3), c(3, 8, 4))
    points(c(1, 2, 3), c(3, 8, 4), pch = 15)
  }
# (3)
  fun3 <- function(){
    plot(c(2, 5), c(-2, 4), xlab = "x", ylab = "y", type = "n",
    main = "Area-3")
    lines(c(2, 4, 5), c(3, -1, 4))
```

```
    points(c(2, 4, 5), c(3, -1, 4), pch = 6)
  }
# (4)
  par(omi = c(0, 0, 0, 0))
# (5)
  layout(matrix(c(0, 1, 3, 2), ncol = 2, byrow = T),
    widths = c(1.6, 1.4), heights = c(5.2, 4.1))
# (6)
  par(mai = c(0.9, 0.7, 0.2, 0.1))
  fun1()
# (4)
  par(mai = c(0.9, 0.7, 0.2, 0.1))
  fun2()
# (5)
  par(mai = c(0.9, 0.7, 0.2, 0.1))
  fun3()
}
```

(1) The function fun1() for drawing the first graph is obtained.
(2) The function fun2() for drawing the second graph is obtained.
(3) The function fun3() for drawing the third graph is obtained.
(4) The graphics area is set.
(5) The command layout() divides the area into areas on a grid. The matrix matrix(c(0, 1, 3, 2), ncol = 2, byrow = T) divides the area into areas on the 2 × 2 grid. Nothing is drawn in the upper left area. The first graph is drawn in the upper right area. The third graph is drawn in the bottom left area. The second graph is drawn in the bottom right area. The argument widths = c(1.6, 1.4) specifies 1.6:1.4 ratio horizontal division. The argument heights = c(5.2, 4.1) sets a 5.2:4.1 ratio vertical division.
(6) After the first area (upper right) is set, the graph is drawn in this area.
(7) After the second area (bottom right) is set, the graph is drawn in this area.
(8) After the third area (bottom left) is set, the graph is drawn in this area.

When the graphics area is divided into two areas horizontally and vertically to locate the four graphs on a grid, adjacent areas can be united. Fig. 1.49 (right) given by Program (1 - 44) is realized.

```
Program (1 - 44)
function() {
# (1)
  fun1 <- function(){
    plot(c(1, 3), c(2, 9), xlab = "x", ylab = "y", type = "n",
      main = "Area-1")
    lines(c(1, 2, 3), c(2, 8, 9))
    points(c(1, 2, 3), c(2, 8, 9), pch = 4)
  }
```

```
# (2)
  fun2 <- function(){
    plot(c(1, 3), c(2, 9), xlab = "x", ylab = "y", type = "n",
    main = "Area-2")
    lines(c(1, 2, 3), c(3, 8, 4))
    points(c(1, 2, 3), c(3, 8, 4), pch = 15)
  }
# (3)
  par(omi = c(0, 0, 0, 0))
# (4)
  layout(matrix(c(0, 1, 2, 2), ncol = 2, byrow = T),
  widths = c(1.6, 1.4), heights = c(5.2, 4.1))
# (5)
  par(mai = c(0.9, 0.7, 0.2, 0.1))
  fun1()
# (6)
  par(mai = c(0.9, 0.7, 0.2, 0.1))
  fun2()
}
```

(1) The function fun1() for drawing the first graph is obtained.
(2) The function fun2() for drawing the second graph is obtained.
(3) The graphics area is set.
(4) The command layout() divides the area into areas on a grid. The matrix
matrix(c(0, 1, 2, 2), ncol = 2, byrow = T) divides the area into 2×2
areas. Nothing is drawn in the upper left area. The first area is drawn in the
upper right area. The second graph is drawn in the area given by uniting the
two areas in the bottom.
(5) After the first area (upper right) is set, the graph is drawn in the area.
(6) After the second area (bottom) is set, the graph is drawn in the area.

The use of par(new = T) is another option for positioning plural graphs
at intended positions in a graphics window. For example, Program (1 -
45) yields Fig. 1.50.

```
Program (1 - 45)
function() {
# (1)
  fun1 <- function(){
    plot(c(1, 3), c(2, 9), xlab = "x", ylab = "y", type = "n")
    lines(c(1, 2, 3), c(2, 8, 9))
    points(c(1, 2, 3), c(2, 8, 9), pch = 14)
  }
# (2)
  fun2 <- function(){
    plot(c(1, 3), c(2, 9), xlab = "x", ylab = "y", type = "n")
    lines(c(1, 2, 3), c(3, 8, 4))
```

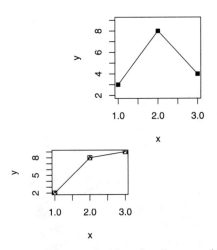

Figure 1.50 Two graphs given by Program (1 - 45).

```
   points(c(1, 2, 3), c(3, 8, 4), pch = 15)
 }
# (3)
 par(mfrow = c(1, 1), omi = c(0, 0, 0, 0))
# (4)
 par(fig = c(0.0, 0.6, 0.0, 0.4), mai = c(0.8, 0.8, 0.1, 0.1))
# (5)
 fun1()
# (6)
 par(new = T)
# (7)
 par(fig = c(0.3, 1, 0.4, 1), mai = c(0.8, 0.8, 0.3, 0.3))
# (8)
 fun2()
}
```

(1) The function **fun1**() for drawing the first graph is obtained.
(2) The function **fun2**() for drawing the second graph is obtained.
(3) The graphics area is set. The argument **omi** = is set in **par**(). However,
mai = is not set.
(4) Specification of **fig** = c(0.0, 0.6, 0.0, 0.4) in **par**() sets the first
area. The argument **mai** = c(0.8, 0.8, 0.1, 0.1) specifies the size of the
figure margins.
(5) The command **par**(**new** = T) describes the superimposition of a new graph
on the current graph.

(6) The first graph is drawn.
(7) The specification of `fig = c(0.3, 1, 0.4, 1)` in `par()` sets the area of the second graph. The argument `mai = c(0.8, 0.8, 0.3, 0.3)` sets the size of the figure margin in the graph.
(8) The second graph is drawn.

Figure 1.51   Bar plot with an image, given by Program (1 - 46).

Since `par(new = T)` enables the superimposition of graphs, a graph can be superimposed on an image displayed by `image()` in black and white; the image is obtained by a digital camera.

For example, Program (1 - 46) yields Fig. 1.51.

```
Program (1 - 46)
function() {
# (1)
   image1 <- read.csv(file = "d:\\GraphicsR\\image1.txt",
   header=F)
# (2)
   par(mai = c(1, 2, 1, 1), omi = c(0, 0, 0, 0))
# (3)
   image1 <- matrix(unlist(image1), nrow = 323)
# (4)
   image(image1, col = gray(seq(from = 0.1, to = 1,
   length = 100)), axes = F)
# (5)
   par(new = T)
# (6)
   yy <- c(225, 500, 641, 192)
```

```
name1 <- c("data-a", "data-b", "data-c", "data-d")
xx <- barplot(yy, ylab = "y", names.arg = name1,
  ylim = c(0, 1000), density = 80, angle = c(30, 60, 90, 120))
}
```

(1) The brightness data file image1.txt is retrieved and named `image1`. For the origin of image1.txt, refer to the Appendix.

(2) The graphics area is set.

(3) `image1` is converted to a matrix of 323×389 size. The command `unlist()` transforms the elements of the matrix into numerical form.

(4) `image1` is shown as an image on the display. The argument `col=gray(seq (from=0.1, to=1, length=100))` makes the image a black-and-white image represented by stepwise thickness. Since `axes = F` is set, no coordinate axes are drawn.

(5) The command `par(new = T)` describes the superimposition of a new graphs on the current graph.

(6) A bar plot is illustrated.

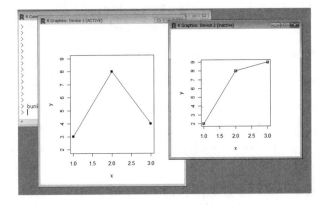

Figure 1.52   Display of graphs in plural graphics window, given by Program (1 - 47).

Plural graphics window for respective graphs can be produced. The command `dev.set(1)` realizes it. Program (1 - 47) achieves Fig. 1.52.

```
Program (1 - 47)
function() {
# (1)
  par(mfrow = c(1, 1), mai = c(1, 1, 1, 1), omi = c(0, 0, 0, 0))
# (2)
  plot(c(1, 3), c(2, 9), xlab = "x", ylab = "y", type = "n")
  lines(c(1, 2, 3), c(2, 8, 9))
  points(c(1, 2, 3), c(2, 8, 9), pch = 14)
# (3)
```

```
  dev.set(1)
# (4)
  par(mfrow = c(1, 1), mai = c(1, 1, 1, 1), omi = c(0, 0, 0, 0))
  plot(c(1, 3), c(2, 9), xlab = "x", ylab = "y", type = "n")
  lines(c(1, 2, 3), c(3, 8, 4))
  points(c(1, 2, 3), c(3, 8, 4), pch = 15)
}
```

(1) The graphics area is set.

(2) A graph is drawn.

(3) The command `dev.set(1)` produces a new graphics window.

(4) The graphics area is set.

(5) A graph is drawn.

1.15 SUMMARY

1. The "work image file" ".RData" starts up R.

   R commands and R programs are executed in a console window.

2. An editor is used for editing

3. The shutdown and rebooting of R reset the settings of graphics.

4. The command `par()` specifies the graphics settings or displays the contents of the settings.

5. The command `plot()` sets the coordinate axes and/or draws the coordinate axes.

6. When the argument `xlab =` is specified in `plot()`, the explanation is written on the x-axis . When `ylab =` is specified, the explanation is written on the y-axis.

7. The first two arguments in `plot()` are the coordinates of the x-axis and y-axis, respectively.

8. If `type = "n"` is set in `plot()`, no data points are drawn.

9. The command `lines()` draws straight lines.

10. The command `points()` draws marks.

11. If `axes = F` is set in `plot()`, no coordinate axes are drawn.

12. Data are stored in variables (a variable is called an object in exact R terminology).

13. The command `text()` writes letters.

14. Mathematical expressions are written using `expression()`.

15. The argument `lwd =` in `lines()` sets the thickness of straight lines.

16. The argument `lty =` set in `lines()` specifies the type of straight lines.

17. The argument `col =` set in `lines()` specifies the color of straight lines.

18. The value of `gray()` specified by `col =` in `lines()` changes the thickness of gray.

19. The value of `pch =` set in `points()` specifies the type of marks.

20. The argument `font =` set in `text()` gives the type of font.

21. The command `symbols()` draws circles, rectangles, stars, thermometers, and boxplots.

22. The command `rect()` draws rectangles.

23. The command `polygon()` draws polygons.

24. `legend()` adds a legend.

25. The argument `log = "y"` set in `plot()` makes the axis the logarithmic scale.

26. The command `barplot()` illustrates a bar plot.

27. The command `pie()` draws a pie chart.

28. The argument `mfrow =` set in `par()` divides a graphics windows into equal-spaced areas.

29. The arguments `mai =` and `omi =` set in `par()` specify the sizes and locations of graphs.

30. The command `layout()` divides the area into areas on a grid.

31. The argument `fig =` set in `par()` specifies the size and location of each graph.

32. The arguments `widths =` and `heights =` set in `layout()` specify the ratio of the division of a graphics window.

33. The command `par(new = T)` describes the superimposition of a new graphs on the current graph.

34. The command `image()` displays a black-and-white image.

35. The `dev.set(1)` produces a new graphics window.

36. The selection of "Copy as metafile" enables copying of a graph to a document of another sprogram.

37. The command `postscript()` constructs a digital file in postscript format.

38. The argument `family` = set in `postscript()` specifies the type of font.

39. The command `jpeg()` constructs a digital file in jpeg format.

40. The command `pdf()` constructs a digital file in pdf format.

EXERCISES

1.1 Replace part (1) of Program (1 - 2)(page 11) with the list below. The position and size of the graph will be altered.

```
par(mai = c(2, 2, 1, 1), omi = c(0.5, 0.5, 0.5, 0.5))
```

1.2 Replace part (1) in Program (1 - 3)(page 13) with the list below. That is, `las = 1` is added to the arguments of `par()`. The values along the y-axis will be positioned horizontally.

```
par(mai = c(1, 1, 1, 1), omi = c(0, 0, 0, 0), las = 1)
```

1.3 Replace part (2) of Program (1 - 4)(page 13) with the list below. The values along the axes will be altered.

```
plot(c(-20, 30), c(25, 95), xlab = "x", ylab = "y")
```

1.4 In part (2) of Program (1 - 5)(page 14), add `main = "Graph-ABC"` to the arguments of `plot()`. That is, replace part (2) with the list below. The title of the graph will appear.

```
plot(c(-2, 3), c(2, 9), xlab = "x", ylab = "y", type = "n",
  main = "Graph-ABC")
```

1.5 Replace parts (3) and (4) in Program (1 - 6)(page 14) with the list below. The positions of the straight lines and small circles will be altered.

```
lines(c(-1, 0.5, 2.5), c(4, 7, 2.5))
points(c(-1, 0.5, 2.5), c(4, 7, 2.5))
```

1.6 Replace part (2) in Program (1 - 7)(page 15) with the list below. The positions of the straight lines and data points will be altered.

```
xx <- c(-1, 0.5, 2.5)
yy <- c(4, 7, 2.5)
```

1.7   Replace part (3) in Program (1 - 14)(page 24) with the list below. The size and position of the text will be changed.

```
points(0.7, 0.8)
text(0.7, 0.8, "World Peace", cex = 1)
```

1.8   The part of (3) of Program (1 - 19)(page 29) is replaced with the list below. `rgb()` constructs colors. `max = 255` indicates that the values of `red =`, `green =`, `blue =`, `alpha =` ranges from 0 to 255. `alpha =` specifies transparency. `alpha = 255` means perfect opacity. However, some R commands gives opaque colors even if translucent colors are specified. Furthermore, even if translucent colors are shown in a display, they may be transformed into perfect opaque colors in digital files.

```
# (3)'
  rgb1 <- rgb(red = 200, green = 50, blue = 100, alpha = 255,
    max = 255)
  text(0, 1, 'col="rgb1"')
  lines(c(0.3, 1), c(1, 1), lwd = 2, col = rgb1)
```

1.9   To make the scale of the x-axis the logarithmic scale, (2) of Program (1 - 29)(page 42) is replaced with the following R program.

```
plot(c(1, 9), c(10, 1000), xlab = "x", ylab = "y", type = "n",
  log = "x")
```

1.10   To make the bars colorful, replace (4) of Program (1 -30)(page 42) with the following command.

```
barplot(yy, ylab = "y", names.arg = name1, col = rainbow(4))
```

1.11   To add a legend, replace (4) of Program (1 - 31)(page 43) with the command below.

```
barplot(yy, xlab = "x", ylab = "y",
  names.arg = c("A-group", "B-group"),
legend = c("q1", "q2", "q3", "q4"))
```

1.12   To make the pie chart colorful, replace (4) of Program (1 - 33)(page 45) with the following command.

```
pie(yy, labels = name1, col = topo.colors(4))
```

1.13   Set `col` = as one of the following specifications; `col` = is the argument
in `image()` in (4) of Program (1 - 46)(page 58).
```
col = rainbow(50)
col = rainbow(30, start = 0.7, end = 0.1)
col = rainbow(2, start = 0.3, end = 0.2)
col = topo.colors(10)
col = terrain.colors(20)
col = heat.colors(3)
col = cm.colors(10)
```

CHAPTER 2

GRAPHICS FOR STATISTICAL ANALYSIS

2.1 INTRODUCTION

The first chapter describes basic techniques in R graphics that enable the construction of R programs, the production of graphs, and the saving of graphs as digital files. This chapter describes methods of drawing more diverse graphs on the basis of the previous chapter. Each R program can be used by itself in the same way as those in Chapter 1. Most of the R graphs in this chapter are not too complex but are of high practical value; they were selected from diverse R graphs.

Furthermore, the author is careful in reducing the number of arguments. This is because most R commands are designed to draw standard graphs when arguments are abbreviated. In addition, laborious works of treating a large number of arguments should not cancel out the beauty of beneficial R commands. However, appropriate settings of arguments improve the power of expression of R commands described in this chapter. A tactful use of arguments for producing a wide variety of graphs is one of the real thrills given by R. Users are expected to construct graphs on a case-by-case basis, while making the most of the literature and articles on the Internet.

Guidebook to R Graphics Using Microsoft Windows,
First Edition. By Kunio Takezawa
Copyright © 2012 John Wiley & Sons, Inc.

2.2 STEM-AND-LEAF DISPLAYS

When a series of numbers are given as data, a stem-and-leaf display is a good tool for recognizing the distribution of the data. Program (2 - 1) exemplifies it.

```
Program (2 - 1)
function ()
{
  xx <- c(1.92, 4.01, 6.51, 1.40, 1.67, 5.27, 1.42, 0.36,
    3.18, 3.67, 7.48, 2.65, 7.86, 10.78, 2.30, 1.29, 0.31, 0.93,
    2.34, 2.53)
  stem(xx)
}
```

The argument of stem() is data (xx in this example). This program yields the result below.

```
The decimal point is 1 digit(s) to the right of the |
  0 | 001111222233344
  0 | 5778
  1 | 1
```

The leftmost number stands for the tens place digit. The numbers on the right of the bars are the ones place digits; these numbers are obtained by rounding off the data to the nearest ones place. Hence, the number of 1s indicates the number of data the ones place digit of which is 1. One group consists of data the ones place digit of which is in the range between 1 and 4. Another group consists of data the ones place digit of which is in the range between 5 and 9.

However, the rounding off by stem() is sometimes unconventional as indicated by the R program below.

```
Program (2 - 2)
function() {
  stem(c(2.25, 2.6499, 3.55))
}
```

The result is shown below.

```
The decimal point is at the |
  2 | 3
  2 | 6
  3 |
  3 | 5
```

If 3.55 is rounded off to one decimal place, the result should be 3.6. The last line of the output above, however, is 3 | 5.

As for the change in the expression of the stem-and-leaf display, Program (2 - 3) exemplifies it.

Program (2 - 3)
```
function ()
{
  xx <- c(1.92, 4.01, 6.51, 1.40, 1.67, 5.27, 1.42, 0.36,
    3.18, 3.67, 7.48, 2.65, 7.86, 10.78, 2.30, 1.29, 0.31, 0.93,
    2.34, 2.53)
  stem(xx, scale = 2)
}
```

The argument scale = 2 is added to stem(). The argument scale = 2 makes the vertical length of a stem-and-leaf display twice as large as that of the display given by scale = 1; the abbreviation of scale = gives the same display as scale = 1. The result is:

```
The decimal point is at the |
  0 | 34934479
  2 | 335627
  4 | 03
  6 | 559
  8 |
 10 | 8
```

Furthermore, if scale = 3 is set as the argument of stem(), the following result is obtained:

```
The decimal point is at the |
  0 | 349
  1 | 34479
  2 | 3356
  3 | 27
  4 | 0
  5 | 3
  6 | 5
  7 | 59
  8 |
  9 |
 10 | 8
```

2.3 HISTOGRAMS AND PROBABILITY DENSITY FUNCTIONS

A typical graph for displaying the distribution of a series of numbers is a histogram. The command hist(), which is contained in R by default, draws a histogram. The use of command hist() requires caution.

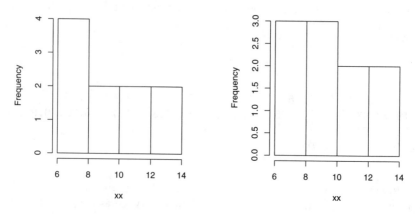

Figure 2.1   Histogram given by   Program (2 - 4) (left).   Histogram   given by
Program (2 - 5) (right).

Fig. 2.1 (left), given by Program (2 - 4), exemplifies it.

Program (2 - 4)
```
function() {
# (1)
 par(mai = c(1, 1, 1, 1), omi = c(0, 0, 0, 0))
# (2)
  xx <- c(9.20, 6.00, 6.00, 11.25, 11.00, 7.25, 9.7, 13.25,
    14.00, 8.00)
# (3)
  hist(xx, breaks = c(6, 8, 10, 12, 14))
}
```

(1) The graphics area is set.
(2) Data (xx) is given.
(3) The command hist() draws a histogram. The argument breaks = c(6,
8, 10, 12, 14) indicates the boundary points (including the minimum point
of the first bin and the maximum point of the last bin) of the bins.
    Among the values of xx, those equal to or larger than 6 and less than 8, are
{6.00, 6.00, 7.25}. However, the first bin in Fig. 2.1 (left) is 4 because the first
bin indicates the number of values equal to or larger than 6, and are equal to
or less than 8. The next bin shows the number of values equal to or larger
than 8, and less than 10. The following bins indicate the number of values
in a similar manner. Hence, Fig. 2.1 (left) does not pass as a conventional
histogam.
    Improvement in this point yields Fig. 2.1 (right) given by Program (2 -
5).

Program (2 - 5)

```
function(){
# (1)
  par(mai = c(1, 1, 1, 1), omi = c(0, 0, 0, 0))
# (2)
      xx <- c(9.20, 6.00, 6.00, 11.25, 11.00, 7.25, 9.7, 13.25,
        14.00, 8.00)
# (3)
  hist(xx, breaks = c(6, 8, 10, 12, 14), right = F)
}
```

In (3), right = F is set in hist(). This is the difference from the previous R program. The frequency of the first bin in Fig. 2.1 (right) is 3. This is the exact number of values equal to or more than 6 and less than 8: $\{6.00, 6.00, 7.25\}$. Other frequencies are defined in the same manner. However, the frequency of the last bin, which should show the number of values equal to or larger than 12 and less than 14, is 2. The value that satisfies this condition is $\{13.25\}$. Hence, the frequency of the last bin is 1. The frequency of the last bin is 2 because $\{14.00\}$ is also counted. Therefore, Fig. 2.1 (right) is not an appropriate histogram.

Then, Program (2 - 6) is attempted.

Program (2 - 6)
```
function() {
# (1)
  par(mai = c(1, 1, 1, 1), omi = c(0, 0, 0, 0))
# (2)
  xx <- c(9.20, 6.00, 6.00, 11.25, 11.00, 7.25, 9.7, 13.25,
    14.00, 8.00)
# (3)
  br1 <-  c(6, 8, 10, 12, 14)
  bw1 <- br1[2] - br1[1]
  xxh <- floor(xx/bw1) * bw1 + 0.1 * bw1
# (4)
  hist(xxh, breaks = br1, right = F)
}
```

(1) The graphics area is set.
(2) Data(xx) is given.
(3) The positions of boundaries of bins are set as br1. The bin width (the horizontal length of the rectangulars in the histogram) is set as bw1. xx is modified for the use of hist(). The result is stored as xxh. The command floor() gives the largest integer, not greater than the elements of the argument. Hence, floor(xx/bw1) in this example is:

4 3 3 5 5 3 4 6 7 4

Therefore, xxh is:

8.2   6.2   6.2 10.2 10.2   6.2   8.2 12.2 14.2   8.2

If the histogram obtained is correct, the bin corresponding to an element of xx is identical to that corresponding to an element of xxh. Furthermore, each value of xxh is slightly larger than that among the values of boundary points. Therefore, the problem described above does not occur because no values of xxh are the same as the values of boundary points.

However, Program (2 - 6) outputs the following:

```
Error in hist.default(xxh, breaks = br1, right = F) :
    some 'x' not counted; maybe 'breaks' do not span range of 'x'
```

Since the range of br1 does not cover that of xxh, this error occurs. This happens because of the fact that xxh contains 14.2.

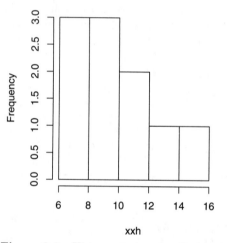

Figure 2.2   Histogram given by Program (2 - 7).

Then, Program (2 - 7) yields Fig. 2.2.

```
Program (2 - 7)
function() {
# (1)
  par(mai = c(1, 1, 1, 1), omi = c(0, 0, 0, 0))
# (2)
  xx <- c(9.20, 6.00, 6.00, 11.25, 11.00, 7.25, 9.7, 13.25,
    14.00, 8.00)
# (3)
  br1 <-  c(6, 8, 10, 12, 14, 16)
  bw1 <- br1[2] - br1[1]
  xxh <- floor(xx/bw1) * bw1 + 0.1 * bw1
```

```
# (4)
  hist(xxh, breaks = br1, right = F)
}
```

In (4), the position of boundary points is set to be c(6, 8, 10, 12, 14, 16). Hence, no errors occur. Furthermore, the resultant histogram (Fig. 2.2) is a correct histogram of xx. Therefore, hist() needs preprocessed data such as that in Program (2 - 7).

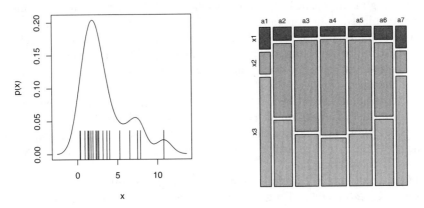

Figure 2.3   Probability density function given by Program (2 - 8) (left). The mosaic plot given by Program (2 - 9) (right).

A histogram is constructed by setting rectangles. However, if the distribution of data is shown by a smooth curve, the characteristics of data are usually represented aptly. Then, Fig. 2.3 (left) exemplifies the representation of data using a smooth curve. This graph is illustrated by Program (2 - 8).

```
Program (2 - 8)
function ()
{
# (1)
  par(mai = c(1, 1, 1, 1), omi = c(0, 0, 0, 0))
# (2)
  xx <- c(1.92, 4.01, 6.51, 1.40, 1.67, 5.27, 1.42, 0.36,
    3.18, 3.67, 7.48, 2.65, 7.86, 10.78, 2.30, 1.29, 0.31, 0.93,
    2.34, 2.53)
# (3)
  d1 <- density(xx, bw = "Sj-ste")
  ex <- d1$x
  ey <- d1$y
# (4)
  plot(ex, ey, type = "n", xlab = "x", ylab = "p(x)")
```

```
   lines(ex, ey)
   rug(xx, ticksize = 0.2, lwd = 1)
}
```

(1) The graphics area is set.
(2) Data (xx) is given.
(3) The command `density()` derives a smooth probability density function using the data. This probability density function is given by a kernel method. Since the argument `kernel =` is not specified in this example, a Gaussian kernel is employed. The constant for setting the smoothness of a probability density function is called a bandwidth. The argument `bw =` in `density()` sets the bandwidth. In this example, `bw = "Sj-ste"` specifies that the Sheather and Jones method is carried out using the solve-the-equation method for optimizing the bandwidth.
(4) The commands `plot()` and `lines()` draw a graph of the probability density function. The command `rug()` shows the positions of the data (the elements of xx) by vertical lines on the x-axis. The argument `ticksize = 0.2` specifies the length of the vertical lines. The argument `lwd = 1` indicates the thickness of the vertical lines.

A mosaic plot enables a simultaneous graphical presentation of values and fractions. Fig. 2.3 (right), Program (2 - 9), presents a simple example.

Program (2 - 9)

```
function() {
# (1)
   par(mai = c(1, 1, 1, 1), omi = c(0, 0, 0, 0))
# (2)
  xx1 <- rep(0.1, length = 7)
  xx2 <- sin(seq(from = 0.1, to = pi-0.1, length = 7))
  xx3 <- rep(0.5, length = 7)
  names(xx1) <- c("a1", "a2", "a3", "a4", "a5", "a6", "a7")
  xx <- cbind(x1 = xx1, x2 = xx2, x3 = xx3)
# (3)
   mosaicplot(xx, col = c("red", "green", "pink"),
     cex.axis = 0.7, main = "")
}
```

(1) The graphics area is set.
(2) Data (xx) is given.
(3) The command `mosaicplot()` illustrates a mosaic plot. The horizontal lengths of rectangles a1, a2, a3, a4, a5, a6, and a7 show the summations of the values of the groups. The division of each group indicates the ratio of the values belonging to the group.

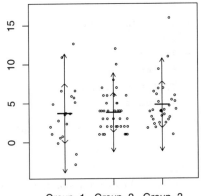

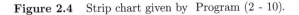

Group–1 Group–2 Group–3

Figure 2.4   Strip chart given by Program (2 - 10).

2.4   STRIP CHART

In comparing the distributions, a strip chart is of great use. Fig. 2.4 given by
Program (2 - 10) exemplifies it.

Program (2 - 10)

```
function ()
{
# (1)
  par(mai = c(1, 1, 1, 1), omi = c(0, 0, 0, 0))
# (2)
  set.seed(591)
  xx1 <- rnorm(20, mean = 3, sd = 3.6)
  xx2 <- rpois(40, lambda = 3.5)
  xx3 <- rchisq(31, df = 5)
# (3)
  mean1 <- c(mean(xx1), mean(xx2), mean(xx3))
  sd1 <- c(sd(xx1), sd(xx2), sd(xx3))
  data1 <- list(xx1, xx2, xx3)
# (4)
  xmin1 <- min(xx1, xx2, xx3) - 1
  xmax1 <- max(xx1, xx2, xx3) + 1
# (5)
  stripchart(data1, method = "jitter", jit = 0.3, vert = T,
    pch = 1, cex = 0.4, ylim = c(xmin1, xmax1),
```

```
    group.names = c("Group-1", "Group-2", "Group-3"))
# (6)
  arrows(1:3, mean1 + sd1, 1:3, mean1 - sd1, angle = 45,
    code = 3, length = 0.07)
  arrows(1:3, mean1 + 2 * sd1, 1:3, mean1 - 2 * sd1,
    angle = 30, code = 3, length = 0.07)
  arrows(1:3, mean1 + 0.01 * sd1, 1:3, mean1 - 0.01 * sd1,
    angle = 90, code = 3, length = 0.12)
}
```

(1) The graphics area is set.

(2) Simulation data is produced. The elements of xx1 are realizations of a normal distribution. In rnorm(), 20 sets the number of data, mean = 3 indicates the mean of the normal distribution, and sd = 3.6 specifies the standard diviation. The elements of xx2 are realizations of a Poisson distribution. In rpois(), 40 represents the number of data, and lambda = 3.5 shows the mean of the Poisson distribution. The elements of xx3 are realizations of a chi-squared distribution. In rchisq(), 31 stands for the number of data, df = 5 represents the degrees of freedom of the chi-squared distribution.

(3) The respective means of xx1, xx2, and xx3 are stored in mean1 as a vector. The respective standard deviation of xx1, xx2, and xx3 are stored in sd1 as a vector. xx1, xx2, and xx3 are put together in a list format using list().

(4) The minimum of the horizontal range of the strip chart is set as xmin1. Its maximum is set as xmax1.

(5) The command stripchart() produces a strip chart. The argument method = "jitter" specifies that the data are plotted at randomly scattered positions in the orthogonal direction of the axis of the data value. The argument jit = 0.3 indicates the variance of the scattering. The argument vert = T sets the axis of the data value horizontally. The argument pch = 1 sets the type of symbol for plotting the data points. The argument cex = 0.4 indicates the size of the symbols. The argument ylim = c(xmin1, xmax1) specifies the range of the vertical axis. The argument group.names = c("Group-1", "Group-2", "Group-3") gives the names of xx1, xx2, and xx3.

(6) The command arrows() adds arrows to the graph. The first argument 1:3 indicates that the values of the x-axis of the starting points of the three arrows are $\{1, 2, 3\}$. mean1 + sd1 represents the starting points of the y-axis of the starting points of the three arrows. The second argument 1:3 means that the values of the x-axis of the starting points of the three arrows are $\{1, 2, 3\}$. mean1 - sd1 represents the ending points of the y-axis of the starting points of the three arrows. The argument angle = 45 gives the angle from the shaft of the arrow to the edge of the arrowhead. The argument code = 3 assigns the type of arrow to be drawn. The argument length = 0.07 represents the length of the edges of the arrowhead (in inches). The first arrows() shows the

range of onefold standard deviation. The second `arrows()` shows the range of twofold standard deviation. The third `arrows()` shows the means.

2.5 BOXPLOTS

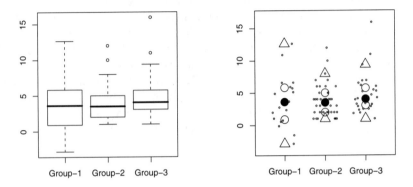

Figure 2.5   Boxplot given by Program (2 - 11) (left). Boxplot given by Program (2 - 12) (right).

The use of a boxplot is another beneficial method of comparing the distribution of data. Fig. 2.5 (left) is an example of a boxplot. In Fig. 2.5 (left), the bottom line of each rectangle is called a lower hinge. The upper line of each rectangle is called an upper hinge. Data is set as $\mathbf{X} = \{X_1, X_2, \cdots, X_n\}(X_1 \leq X_2 \leq \cdots \leq X_n)$. The lower hinge is represented as $l(\mathbf{X})$. The upper hinge is represented as $u(\mathbf{X})$. Then, $l(\mathbf{X})$ and $u(\mathbf{X})$ are calculated as

$$
\begin{cases}
n \text{ is odd} & \begin{aligned}
p &= \text{ceiling}(0.25 \cdot (n+3)), & l_a &= X_p \\
q &= \text{floor}(0.25 \cdot (n+3)), & l_b &= X_q \\
l(\mathbf{X}) &= (l_a + l_b) \cdot 0.5
\end{aligned} \\
n \text{ is even} & \begin{aligned}
p &= \text{ceiling}(0.25 \cdot (n+2)), & l_a &= X_p \\
q &= \text{floor}(0.25 \cdot (n+2)), & l_b &= X_q \\
l(\mathbf{X}) &= (l_a + l_b) \cdot 0.5
\end{aligned}
\end{cases}
\tag{2.1}
$$

$$
\begin{cases}
n \text{ is odd} & \begin{aligned}
p &= n+1 - \text{ceiling}(0.25 \cdot (n+3)), & u_a &= X_p \\
q &= n+1 - \text{floor}(0.25 \cdot (n+3)), & u_b &= X_q \\
u(\mathbf{X}) &= (u_a + u_b) \cdot 0.5
\end{aligned} \\
n \text{ is even} & \begin{aligned}
p &= n+1 - \text{ceiling}(0.25 \cdot (n+2)), & u_a &= X_p \\
q &= n+1 - \text{floor}(0.25 \cdot (n+2)), & u_b &= X_q \\
u(\mathbf{X}) &= (u_a + u_b) \cdot 0.5
\end{aligned}
\end{cases}
\tag{2.2}
$$

median() means a median; the R command `median()` plays this role. The command ceiling() stands for the smallest integer not less than the values of the argument; the R command `ceiling()`) plays this role. The command floor() gives the largest integer, not greater than the values of the argument; the R command `floor()` plays this role. Furthermore, the quantile region of **X** ($w(\mathbf{X})$) is defined as;

$$w(\mathbf{X}) = u(\mathbf{X}) - l(\mathbf{X}). \tag{2.3}$$

In Fig. 2.5 (left), the downward line from each rectangle is called a lower whisker. The upward line from each rectangle is called an upper whisker. The end of a lower whisker is represented as $lw(\mathbf{X})$. The end of an upper whisker is represented as $uw(\mathbf{X})$. $lw(\mathbf{X})$ is the minimum value of the selected elements of **X**; the selected elements satisfy the condition that the values of the elements are equal to or larger than ($l(\mathbf{X}) - 1.5lw(\mathbf{X})$). $uw(\mathbf{X})$ is the maximum value of the selected elements of **X**; the selected elements satisfy the condition that the values of the elements are equal to or less than ($u(\mathbf{X}) + 1.5lw(\mathbf{X})$).

Fig. 2.5 (left) given by Program (2 - 11) is an example.

```
Program (2 - 11)
function ()
{
# (1)
  par(mai = c(1, 1, 1, 1), omi = c(0, 0, 0, 0))
# (2)
  set.seed(591)
  xx1 <- rnorm(20, mean = 3, sd = 3.6)
  xx2 <- rpois(40, lambda = 3.5)
  xx3 <- rchisq(31, df = 5, ncp = 0)
# (3)
  box1 <- boxplot(xx1, xx2, xx3, names = c("Group-1", "Group-2",
   "Group-3"), cex = 0.7)
# (4)
  print("box1$stats")
  print(box1$stats)
}
```

(1) The graphics area is set.
(2) Simulation data is produced. The elements of **xx1** are realizations of a normal distribution. The elements of **xx2** are realizations of a Poisson distribution. The elements of **xx3** are realizations of a chi-squared distribution.
(3) The command `boxplot()` draws a boxplot. The argument names = c("Group-1", "Group-2", "Group-3") gives the names of the three data sets represented along the three axes. The argument cex = 0.7 sets the size of the data points plotted below the lower whisker and above the upper whisker.
(4) The values show the positions of the lower whiskers ($lw(\mathbf{X})$), lower hinges ($l(\mathbf{X})$), the medians (median($\mathbf{X}$)), the upper whiskers ($u(\mathbf{X})$), and the upper whiskers ($uw(\mathbf{X})$). This part outputs the following:

```
[1] "box1$stats"
          [,1] [,2]        [,3]
[1,] -2.8580738  1.0 0.9967114
[2,]  0.8986256  2.0 3.0599321
[3,]  3.6043115  3.5 4.0441146
[4,]  5.8154664  5.0 5.7286438
[5,] 12.6367499  8.0 9.3837688
```

Three columns represent three groups of data.

Using Eq.(2.1), Eq.(2.2), and Eq.(2.3), a graph similar to that shown in Fig. 2.5 (left) is drawn without using `boxplot()`. Fig. 2.5 (right) given by Program (2 - 12) exemplifies it.

```
Program (2 - 12)
function ()
{
# (1)
  par(mai = c(1, 1, 1, 1), omi = c(0, 0, 0, 0))
# (2)
  hlow <- function(xx){
    xx <- sort(xx)
    nd <- length(xx)
    mod1 <- nd %% 2
    if(mod1 == 1){
      np <- ceiling(0.25 * (nd + 3))
      nq <- floor(0.25 * (nd + 3))
      low1 <- 0.5*(xx[np] + xx[nq])
    }
    if(mod1 == 0){
      np <- ceiling(0.25 * (nd + 2))
      nq  <- floor(0.25 * (nd + 2))
      low1 <- 0.5*(xx[np] + xx[nq])
    }
    return(low1)
  }
# (3)
  hup <- function(xx){
    xx <- sort(xx)
    nd <- length(xx)
    mod1 <- nd %% 2
    if(mod1 == 1){
      np <- nd + 1 - ceiling(0.25 * (nd + 3))
      nq <- nd + 1 - floor(0.25 * (nd + 3))
      up1 <- 0.5 * (xx[np] + xx[nq])
    }
    if(mod1 == 0){
```

```
      np <- nd + 1 - ceiling(0.25 * (nd + 2))
      nq <- nd + 1 - floor(0.25 * (nd + 2))
      up1 <- 0.5 * (xx[np] + xx[nq])
    }
    return(up1)
  }
# (4)
  set.seed(591)
  xx1 <- rnorm(20, mean = 3, sd = 3.6)
  xx2 <- rpois(40, lambda = 3.5)
  xx3 <- rchisq(31, df = 5)
# (5)
  med1 <- median(xx1)
  hlow1 <- hlow(xx1)
  hup1 <- hup(xx1)
  wid1 <-  (hup1 - hlow1) * 1.5
  llow1 <-  min(xx1[xx1 >= hlow1 - wid1])
  lup1 <-  max(xx1[xx1 <= hup1 + wid1])
# (6)
  print("Group-1")
  print(llow1)
  print(hlow1)
  print(med1)
  print(hup1)
  print(lup1)
# (7)
  med2 <- median(xx2)
  hlow2 <- hlow(xx2)
  hup2 <- hup(xx2)
  wid2 <-  (hup2 - hlow2) * 1.5
  llow2 <-  min(xx2[xx2 >= hlow2 - wid2])
  lup2 <-  max(xx2[xx2 <= hup2 + wid2])
  print("Group-2")
  print(llow2)
  print(hlow2)
  print(med2)
  print(hup2)
  print(lup2)
  med3 <- median(xx3)
  hlow3 <- hlow(xx3)
  hup3 <- hup(xx3)
  wid3 <-  (hup3 - hlow3) * 1.5
  llow3 <-  min(xx3[xx3 > hlow2 - wid3])
  lup3 <-  max(xx3[xx3 < hup3 + wid3])
  print("Group-3")
```

```
    print(llow3)
    print(hlow3)
    print(med3)
    print(hup3)
    print(lup3)
# (8)
    xmin1 <- min(xx1, xx2, xx3) - 1
    xmax1 <- max(xx1, xx2, xx3) + 1
    data1 <- list(xx1, xx2, xx3)
# (9)
    stripchart(data1, "jitter", jit = 0.3, vert = T, pch = 1,
      cex = 0.4, ylim = c(xmin1, xmax1),
     group.names = c("Group-1", "Group-2", "Group-3"))
    points(1, med1, pch = 16, cex = 2)
    points(1, hlow1, pch = 1, cex = 2)
    points(1, hup1, pch = 1, cex = 2)
    points(1, llow1, pch = 2, cex = 2)
    points(1, lup1, pch = 2, cex = 2)
    points(2, med2, pch = 16, cex = 2)
    points(2, hlow2, pch = 1, cex = 2)
    points(2, hup2, pch = 1, cex = 2)
    points(2, llow2, pch = 2, cex = 2)
    points(2, lup2, pch = 2, cex = 2)
    points(3, med3, pch = 16, cex = 2)
    points(3, hlow3, pch = 1, cex = 2)
    points(3, hup3, pch = 1, cex = 2)
    points(3, llow3, pch = 2, cex = 2)
    points(3, lup3, pch = 2, cex = 2)
}
```

(1) The graphics area is set.

(2) The function `hlow()` for deriving the position of a lower hinge is defined using Eq.(2.1) (page 75).

(3) The function `hup()` for deriving the position of an upper hinge is defined using Eq.(2.2) (page 75).

(4) Simulation data is produced. The elements of xx1 are realizations of a normal distribution. The elements of xx2 are realizations of a Poisson distribution. The elements of xx3 are realizations of a chi-squared distribution.

(5) The values of med1 (median, hlow1 (the position of the lower hinge), hup1 (the position of the upper hinge), llow1 (the position of the lower whisker), and lup1 (the position of the upper whisker) are calculated.

(6) The values obtained in (5) are displayed.

(7) The steps of (5) and (6) are conducted using xx2 and xx3.

(8) xx1, xx2, and xx3 are put together in a list format using list(). The minimum of the horizontal range of the strip chart is set as xmin1; its maximum is set as xmax1.

(9) The command stripchart() produces a strip chart. The values derived in (5) and (7) are plotted in the graph.

This R program outputs the following numbers, which are identical to those given by Program (2 - 11).

```
[1] "Group-1"
[1] -2.858074
[1] 0.8986256
[1] 3.604311
[1] 5.815466
[1] 12.63675
[1] "Group-2"
[1] 1
[1] 2
[1] 3.5
[1] 5
[1] 8
[1] "Group-3"
[1] 0.9967114
[1] 3.059932
[1] 4.044115
[1] 5.728644
[1] 9.383769
```

2.6 MULTIPLE-AXIS LAYOUTS

Plural axes are required in some cases to compare the data. Fig. 2.6 (left) given by Program (2 - 13) exemplifies it.

```
Program (2 - 13)
function() {
# (1)
  par(mai = c(1, 1, 1, 1), omi = c(0, 0, 0, 0))
# (2)
  plot(c(-0.1, 2.1), c(0, 2.3), type = "n", xlab = "x",
   ylab = "y")
  xx <- c(0, 1,2)
  yy <- c(2, 0.8, 1.4)
  lines(xx, yy)
  points(xx, yy, pch = 0)
# (3)
  text(xx, yy + 0.2, labels = as.character(yy))
```

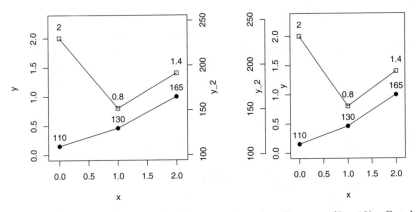

Figure 2.6   Graph with plural y-axes, given by Program (2 - 13). Graph with plural y-axes, given by Program (2 - 14).

```
# (4)
  par(new = T)
# (5)
  plot(c(-0.1, 2.1), c(100, 250), type = "n", xlab = "",
  ylab = "", axes = F)
# (6)
  axis(4)
  mtext("y_2", side = 4, line = 2)
# (7)
  xx <- c(0, 1, 2)
  yy <- c(110, 130, 165)
  lines(xx, yy)
  points(xx, yy, pch = 16)
# (8)
  text(xx, yy + 10, labels = as.character(yy))
}
```

(1) The graphics area is set.
(2) A graph of the first data set is drawn.
(3) The values of the first data set are written in the graph.
(4) The command `par(new = T)` describes that the current graph will remain as it is and a new graph will be superimposed.
(6) The command `axis(4)` draws a second y-axis on the right-hand side of the graph. The command `mtext()` adds an explanation (`y_2`) of the second y-axis. The argument `side = 4` indicates that the y-axis is drawn on the right-hand

side of the graph. The argument `line = 2` sets the distance between the axis and an explanation of the axis.

(7) The second data set is drawn.

(8) The values of the second data set are written in the graph.

   Another y-axis can be added on the left-hand side of a graph. Fig. 2.6 (right) given by Program (2 - 14) is an example.

Program (2 - 14)

```
function() {
# (1)
  par(mai = c(1, 2, 1, 1), omi = c(0, 0, 0, 0))
# (2)
  plot(c(-0.1, 2.1), c(0, 2.3), type = "n",
   xlab = "x", ylab = "")
  mtext("y", side = 2, line = 0.3)
  xx <- c(0, 1, 2)
  yy <- c(2, 0.8, 1.4)
  lines(xx, yy)
  points(xx, yy, pch = 0)
# (3)
  text(xx, yy + 0.2, labels = as.character(yy))
# (4)
  par(new = T)
# (5)
  plot(c(-0.1, 2.1), c(100, 250), type = "n", xlab = "x",
   ylab = "", axes = F)
  xx <- c(0, 1, 2)
  yy <- c(110, 130, 165)
  lines(xx, yy)
  points(xx, yy, pch = 16)
# (6)
  text(xx, yy + 10, labels = as.character(yy))
# (7)
  par(new = T)
# (8)
  par(mai = c(1, 1, 1, 1))
# (9)
  axis(2)
  mtext("y_2", side = 2, line = 0.3)
}
```

(1) The graphics area is set.

(2) The graph of the first data set is drawn. The command `mtext()` adds an explanation (y) of the first y-axis.

(3) The values of the first data set are written in the graph.

(4) The command **par(new = T)** describes that the current graph will remain as it is and a new graph will be superimposed.
(5) The second data set is drawn.
(6) The values of the second data set are written in the graph.
(7) The command **par(new = T)** describes that the current graph will remain as it is and a new graph will be superimposed.
(8) The graphics area is adjusted. Since **mai = c(1, 2, 1, 1)** is substituted with **mai = c(1, 1, 1, 1)**, the figure margin on the left-hand side of the graph decreases in width.
(9) The command **axis(2)** adds a y-axis on the left-hand side of the graph. The command **mtext()** adds an explanation (**y_2**) of the second y-axis.

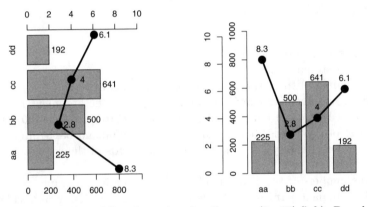

Figure 2.7   Bar plot and line chart given by Program (2 - 15) (left). Bar plot and line chart given by Program (2 - 16) (right).

Both a bar plot and a line chart are drawn in a graph. Fig. 2.7 (left) given by Program (2 - 15) exemplifies it.

Program (2 - 15)
```
function() {
# (1)
  par(mai = c(1, 2, 1, 1), omi = c(0, 0, 0, 0))
# (2)
  xx <- c(225, 500, 641, 192)
# (3)
  name1 <- c("aa", "bb", "cc", "dd")
# (4)
  yy <- barplot(xx, ylab = "", names.arg = name1, horiz = T,
    xlim = c(0, 1000))
# (5)
  text(xx + 80, yy,  labels = as.character(xx))
# (6)
```

```
  par(new = T)
# (7)
  xx2 <- c(8.3, 2.8, 4, 6.1)
# (8)
  plot(xx2, yy, xlim = c(0.38, 10), axes = F, type = "n",
   xlab = "", ylab = "")
# (9)
  lines(xx2 , yy, lwd = 3)
  points(xx2 , yy, pch = 16, cex = 2)
# (10)
  text(xx2 + 1, yy, labels = as.character(xx2))
# (11)
  axis(3)
}
```

(1) The graphics area is set.
(2) Data is stored as **xx**.
(3) The names of the respective data are set as **name1**.
(4) The command **barplot()** draws a bar plot. The argument **horiz = T** indicates a horizontal bar plot. The argument **xlim = c(0, 1000)** specifies the range of the y-axis (horizontal axis). The command **barplot()** outputs **yy**. The elements of **yy** are the values of the y-axis of the respective bars of the bar plot.
(5) The numbers of the elements of **xx** are written in the graph.
(6) The command **par(new = T)** describes that the current graph will remain as it is and a new graph will be superimposed.
(7) **xx2** is set as data for a line chart.
(8) New coordinate axes are set. Since the positions of the scale of the x-axis in the bar plot is different from that of the line chat, **xlim = c(0.38, 10)** is set to make the minimum of the x-axis in the line chart roughly 0.
(9) A line chart for **xx2** is drawn.
(10) The numbers of the elements of **xx2** are written in the graph.
(11) A new coordinate axis is drawn on the top side of the graph.

When a bar plot and a line chart are illustrated in a graph, two coordinate axes along the y-axis can be drawn. Fig. 2.7 (right) given by Program (2 - 16) is an example.

```
Program (2 - 16)
function() {
# (1)
  par(mai = c(1, 2, 1, 1), omi = c(0, 0, 0, 0))
# (2)
  yy <- c(225, 500, 641, 192)
# (3)
  name1 <- c("aa", "bb", "cc", "dd")
# (4)
```

```
  xx <- barplot(yy, ylab = "", names.arg = name1,
    ylim = c(0, 1000))
# (5)
  text(xx, yy + 30, labels = as.character(yy))
# (6)
  par(new = T)
# (7)
  xx2 <- c(1,2,3,4)
  yy2 <- c(8.3, 2.8, 4, 6.1)
# (8)
  plot(xx2, yy2, xlim = c(0.6, 4.4), ylim = c(0.35, 10),
    type = "n", axes = F, xlab = "", ylab = "")
  lines(xx2, yy2, lwd = 3)
  points(xx2, yy2, pch = 16, cex = 2)
# (9)
  text(xx2, yy2 + 0.8, labels = as.character(yy2))
# (10)
  par(new = T)
# (11)
  par(mai = c(1, 1, 1, 1))
# (12)
  axis(2)
}
```

(1) The graphics area is set.

(2) Data is set as **yy**.

(3) The names of the respective data are set as **name1**.

(4) The command **barplot()** draws a bar plot. The argument **xlim** = c(0, 1000) specifies the range of the y-axis (horizontal axis). The command **barplot()** outputs **yy**. The elements of **yy** are the values of the y-axis of respective bars of the bar plot.

(5) The numbers of the elements of **yy** are written in the graph.

(6) The command **par(new = T)** describes that the current graph will remain as it is and a new graph will be superimposed.

(7) **xx2** and **yy2** are provided for drawing a line chart.

(8) A line chart is illustrated. The argument **xlim = c(0.6, 4.4)** is specified for matching the positions of the y-axes of the bar plot and line chart. Since the settings of y-axis are different between the bar plot and the line chart, **ylim = c(0.35, 10)** adjusts it.

(9) The numbers of the elements of **yy2** are written in the graph.

(10) The command **par(new = T)** describes that the current graph will remain as it is and a new graph will be superimposed.

(11) The second graphics area is set.

(12) Second coordinate axes are drawn.

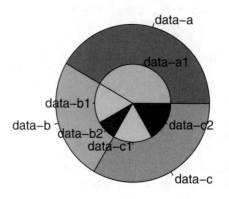

Figure 2.8 Dual-pie chart given by Program (2 - 17).

The technique of adding a new graph on a current graph enables the use of a graph for comparison of two pie charts. Fig. 2.8 given by Program (2 - 17) provides an example.

Program (2 - 17)
```
function() {
# (1)
  par(mai = c(1, 1, 1, 1), omi = c(0, 0, 0, 0))
# (2)
  yy <- c(50,30,40)
# (3)
  name1 <- c("data-a", "data-b", "data-c")
# (4)
  pie(yy, labels = name1, col = c("red","green","skyblue"))
# (5)
  par(new = T)
# (6)
  par(mai=c(2, 2, 2, 2))
# (7)
  yy2 <- c(50, 20, 10, 20, 20)
# (8)
  name2 <- c("data-a1", "data-b1", "data-b2", "data-c1",
   "data-c2")
# (9)
  pie(yy2, labels = name2, col = c("pink", "gold", "blue",
   "gold", "blue"))
}
```

(1) The graphics area is set.
(2) The first data (yy) is given.

(3) The names of the respective elements of the first data are set as `name1`.
(4) The command `pie()` draws a pie chart; `yy` is used as data and `name1` is used as a name of each element of the data.
(5) The command `par(new = T)` describes that the current graph will remain as it is and a new graph will be superimposed.
(6) Since `mai = c(2, 2, 2, 2)` is set in `par()`, the position of the center of the graph will remain as it is and the area for a graph is narrowed.
(7) The second data (`yy2`) is provided.
(8) The names of the respective elements of the second data are set as `name2`.
(9) The command `pie()` draws a pie chart; `yy2` is used as data and `name2` is used as a name of each element of the data.

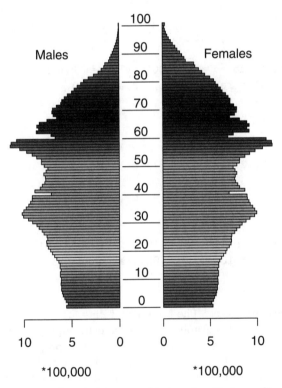

Figure 2.9 Population pyramid given by Program (2 - 18).

An application of a bar plot is a population pyramid. Fig. 2.9 produced by Program (2 - 18) is an example.

Program (2 - 18)
```
function() {
# (1)
  par(omi = c(0, 0, 0, 0), fin = c(5.6, 5.6))
```

```
# (2)
  yy <- read.csv(file = "d:\\GraphicsR\\pop1.csv", header = F)
  yy1 <- yy[,2]
  yy2 <- yy[,3]
# (3)
  par(mai = c(1, 0.5, 0.5, 3.1))
# (4)
  barplot(-yy1, xlab = "*100,000", horiz = T, axes = F,
   col = rainbow(101), space = 0)
  axis(1, labels = c(0, 5, 10), at = c(0, -500, -1000))
  text(-700, 90, "Males")
# (5)
  par(new = T)
  par(mai = c(1, 3.1, 0.5, 0.5))
# (6)
  barplot(yy2, xlab = "*100,000", ylab = "",  horiz = T,
   axes = F, col = rainbow(101), space = 0)
  axis(1, labels = c(0, 5, 10), at = c(0, 500, 1000))
  text(700, 90, "Females")
# (7)
  par(new = T)
  par(mai = c(1, 2.6, 0.5, 2.6))
# (8)
  plot(c(0, 1), c(0, 1), type = "n", xlab = "", ylab = "",
   axes = F)
  for(ii in 0:10){
    lines(c(0,1), c(ii * 10 / 101, ii * 10 / 101))
    text(0.5, ii * 10/101 + 0.025, as.character(ii * 10))
  }
}
```

(1) The graphics area is set. The argument `fin = c(5.6, 5.6)` sets the size of the graphics area. This graph is constructed by superposing three graphs. Hence, if the size of the graphics area is not set appropriately, the desired result is not achieved. This is the reason for the above-described argument. If a graphics window is present and is too small, or a graphics window is not present and the space for a graphics window is too small, the following error message is displayed.

`Error in plot.new() : figure region too large`

If this error occurs, the values in `fin = c(5.6, 5.6)` should be decreased. According to this setting, the values of `mai =` in (3), (5), and (7) should be adjusted. Furthermore, if the display of a personal computer is large so that the size of the population pyramid is larger, the values in `fin = c(5.6, 5.6)` should be changed. Even then, the values of `mai =` in (3), (5), and (7) are also adjusted.

(2) Data is retrieved. Male data is named `yy1`. Female data is named `yy2`. For the origin of jinkou1.csv, refer to Appendix.

(3) A graphics area for depicting the male data (`yy1`) is set.

(4) The command `barplot()` draws the male data (`yy1`). Since the bar chart for male data should extend to the left-hand side, the data is used as `-yy1`. This `barplot()` does not draw axes. The command `axis()` draws axes instead. In this `axis()`, at = `c(0, 500, 1000)` indicates the location for writing letters.

(5) A graphics area for depicting the female data (`yy2`) is set.

(6) The command `barplot()` draws the female data (`yy2`).

(7) The area for displaying age intervals is set at the center.

(8) The horizontal lines for classifying males and females into 10-year age groups are drawn. The ages are written on the lines.

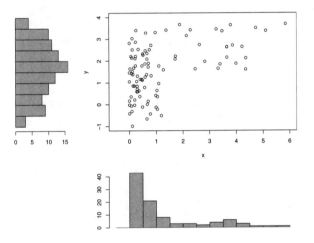

Figure 2.10 Scatter plot with distributions of variables, given by Program (2 - 19).

If the distribution of data with two variables is illustrated in a scatter plot and the distributions of respective variables are represented as two histograms to be included in the graph, the distribution of data with two variables is compared with that with respective variables. Fig. 2.10 given by Program (2 - 19) was drawn considering this line of thought.

Program (2 - 19)
```
function() {
# (1)
  par(omi = c(0.5, 0.5, 0.5, 0.5))
# (2)
  set.seed(349)
```

```
xx <- rnorm(100, mean = 0.5, sd = 1)^2
yy <- rnorm(100, mean = 2, sd = 2) * 0.5 + xx * 0.5
xx1 <- xx
yy1 <- yy
xx <- xx1[-0.5 < xx1 & xx1 < 6 & -1 < yy1 & yy1 < 4]
yy <- yy1[-0.5 < xx1 & xx1 < 6 & -1 < yy1 & yy1 < 4]
# (3)
brx <- seq(from = -0.5, to = 6, by = 0.5)
bwx <- brx[2] - brx[1]
xxh <- floor(xx/bwx) * bwx + 0.1 * bwx
bry <- seq(from = -1, to = 4, by = 0.5)
bwy <- bry[2] - bry[1]
yyh <- floor(yy/bwy) * bwy + 0.1 * bwy
countsx <- hist(xxh, breaks = brx, plot = F)$counts
countsy <- hist(yyh, breaks = bry, plot = F)$counts
# (4)
layout(matrix(c(3, 1, 0, 2), ncol = 2, byrow = T),
  widths = c(1, 3), heights = c(3, 1))
# (5)
par(mai=c(1, 1, 0.1, 0.1))
plot(xx, yy, xlim = range(brx), ylim = range(bry),
  xlab = "x", ylab = "y")
# (6)
par(mai=c(0, 1, 0, 0.1))
barplot(countsx, ylim = c(0, max(countsx)), space = 0)
# (7)
par(mai = c(1, 0.6, 0.1, 0))
barplot(countsy, xlim = c(0, max(countsy)), space = 0,
  horiz = TRUE)
}
```

(1) The graphics area is set.

(2) Simulation data with two variables is produced, the variables are named xx and yy. Elements larger than -0.5 or less than 6 are omitted from xx. Elements larger than -1 or less than 4 are omitted from yy.

(3) The technique for illustrating Fig. 2.2(page 70) is employed to obtain the frequency distribution of xx; the frequency distribution is named countsx. The frequency distribution of yy is derived and named countsy.

(4) The command layout() divides the graphics window. The argument matrix(c(3, 1, 0, 2), ncol = 2, byrow = T) specifies that the area is divided as 2×2 (divided into four areas). The third graph is drawn in the upper left area. The first graph is drawn in the upper right area. The second graph is drawn in the bottom right area. The argument widths = c(1, 3) specifies that the area is divided horizontally as 1:3 when the area is divided

into four areas. The argument `heights = c(3, 1)` specifies that the area is divided vertically as 3:1.

(5) After the first area is set, the scatter plot between `xx` and `yy` is depicted in this area.

(6) After the second area is set, `barplot()` draws the histogram of `xx` in this area. When the second area is set, the scale of the x-axis of the scatter plot should be coincident with that of the histogram.

(7) After the second area is set, `barplot()` draws the histogram of `yy` in this area. When the third area is set, the scale of the y-axis of the scatter plot should be coincident with that of the histogram.

2.7 DISPLAY OF CONFIDENCE INTERVALS

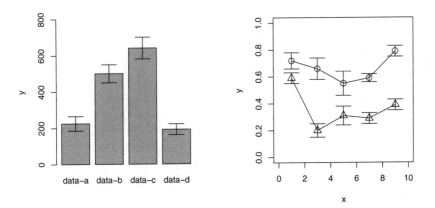

Figure 2.11   Bar plot with confidence intervals, given by Program (2 - 20) (left). Line chart with confidence intervals, given by Program (2 - 21) (right).

Confidence intervals of respective data points should be included in a graph on occasion. For example, Fig. 2.11 (left) given by Program (2 - 20) plays this role.

```
Program (2 - 20)
function() {
# (1)
  par(mai = c(1, 1, 1, 1), omi = c(0, 0, 0, 0))
# (2)
  yy <- c(225, 500, 641, 192)
# (3)
  name1 <- c("data-a", "data-b", "data-c", "data-d")
```

```
# (4)
  xp <- barplot(yy,  ylab = "y", names.arg = name1,
   ylim = c(0, 800))
# (5)
  sd1 <- c(40, 50, 60, 30)
  arrows(xp, yy - sd1, xp, yy + sd1, angle = 90, length = 0.15)
  arrows(xp, yy + sd1, xp, yy - sd1, angle = 90, length = 0.15)
}
```

(1) The graphics area is set.
(2) Data is set as yy.
(3) The names of the respective data are set as name1.
(4) The command barplot() draws a bar plot.
(5) The confidence intervals for the four data points are given as sd1. The command arrows() draws the confidence intervals. The argument angle = 90 indicates that the edge of the arrowhead is flat. The argument length = 0.15 sets the size of the arrowhead.

Confidence intervals can be added to a line chart. Fig. 2.11 (right) given by Program (2 - 21) exemplifies it.

Program (2 - 21)
```
function() {
# (1)
  par(mai=c(1, 1, 1, 1), omi=c(0, 0, 0, 0))
# (2)
  plot(c(0, 10), c(0, 1), xlab = "x", ylab = "y", type = "n")
# (3)
  xx <- c(1, 3, 5, 7, 9)
  yy1 <- c(0.72, 0.66, 0.55, 0.59, 0.79)
  yy2 <- c(0.59, 0.2, 0.31, 0.29, 0.39)
# (4)
  points(xx, yy1, pch = 1, cex = 1.5)
  lines(xx, yy1)
# (5)
  sd1 <- c(0.06, 0.08, 0.09, 0.03, 0.04)
  arrows(xx, yy1 - sd1, xx, yy1 + sd1, angle = 90, length = 0.15)
  arrows(xx, yy1 + sd1, xx, yy1 - sd1, angle = 90, length = 0.15)
# (6)
  points(xx, yy2, pch = 2, cex = 1.5)
  lines(xx, yy2)
# (7)
  sd2 <- c(0.04, 0.05, 0.07, 0.04, 0.04)
  arrows(xx, yy2 - sd2, xx, yy2 + sd2, angle = 90, length = 0.15)
  arrows(xx, yy2 + sd2, xx, yy2 - sd2, angle = 90, length = 0.15)
}
```

(1) The graphics area is set.

(2) The coordinate axes are drawn.

(3) Data is given. Both values of the x-axis of the two data sets are **xx**. The value of the y-axis of the first data set is **yy1**. The value of the y-axis of the second data set is **yy2**.

(4) A line chart is depicted using **xx** and **yy1**.

(5) Confidence intervals are added to the graph drawn in (4).

(6) A line chart is depicted using **xx** and **yy2**.

(7) Confidence intervals are added to the graph drawn in (6).

2.8   SCATTER PLOT MATRICES

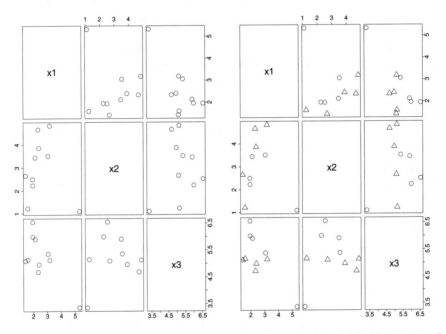

Figure 2.12   Matrix of scatter plots given by Program (2 - 22) (left). Matrix of scatter plots given by Program (2 - 23) (right).

A matrix of scatter plots is available for viewing the relationships between plural variables. Fig. 2.12 (left) given by Program (2 - 22) is an example.

```
Program (2 - 22)
function() {
# (1)
  par(mai = c(1, 1, 1, 1), omi = c(0, 0, 0, 0))
# (2)
  set.seed(35)
```

```
  xx1 <- rnorm(10, mean = 2, sd = 1)
  xx2 <- rnorm(10, mean = 3, sd = 1)
  xx3 <- rnorm(10, mean = 4, sd = 2)
# (3)
  data1 <- data.frame(x1 = xx1, x2 = xx2, x3 = xx3)
# (4)
  pairs(data1, cex.labels = 3, cex.axis = 2, cex = 3)
}
```

(1) The graphics area is set.

(2) Simulation data (xx1, xx2, xx3) is produced. xx1, xx2, and xx3 are realizations of a normal distribution.

(3) xx1, xx2, and xx3 are put together in the data frame data1. xx1, xx2, and xx3 are called x1, x2, and x3, respectively, in data1.

(4) The command pairs() draws a matrix of scatter plots The argument cex.labels = 3 specifies the sizes of "$x1$", "$x2$", and "$x3$" located diagonally to the matrix of scatter plots. The argument cex.axis = 2 sets the size of the letters of the numbers along the coordinate axes. The argument cex = 3 indicates the size of the symbols of data points.

The symbols of the data points in the matrix of scatter plots can differ among the data groups. Fig. 2.12 (right) given by Program (2 - 23) is an example graph.

Program (2 - 23)
```
function() {
# (1)
  par(mai = c(1, 1, 1, 1), omi = c(0, 0, 0, 0))
# (2)
  set.seed(35)
  xx1 <- rnorm(10, mean = 2, sd = 1)
  xx2 <- rnorm(10, mean = 3, sd = 1)
  xx3 <- rnorm(10, mean = 4, sd = 2)
# (3)
  names(xx1) <- c(rep("aaa", 5), rep("bbb", 5))
  pch1 <- rep(1, length = 10)
  pch1[names(xx1) == "bbb"] <- 2
# (4)
  data1 <- data.frame(x1 = xx1, x2 = xx2, x3 = xx3)
# (5)
  pairs(data1, pch = pch1, cex.labels = 3, cex.axis = 2, cex = 3)
}
```

(1) The graphics area is set.

(2) Simulation data (xx1, xx2, xx3) is produced.

(3) The first five elements of xx1 are named **aaa**. The last five elements are named bbb. Using these names, the first five elements of pch1 are set as 1. The last five elements of pch1 are set as 2.

(4) xx1, xx2, and xx3 are put together in the data frame data1.
(5) The command pairs() draws a matrix of scatter plots. Because of the argument pch = pch1, the symbols for the first five data are designated by pch = 1, and those for the last five data are designated by pch = 2.

2.9   RADAR CHARTS AND PARALLEL CHARTS

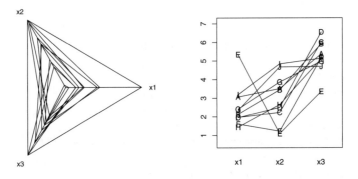

Figure 2.13   Radar chart given by Program (2 - 24) (left). Parallel chart given by Program (2 - 25) (right).

A radar chart is another means of depicting the relationship between data with plural variables. Fig. 2.13 (left) given by Program (2 - 24) exemplifies this.

```
Program (2 - 24)
function() {
# (1)
  par(mai = c(2, 2, 2, 2), omi = c(1, 1, 1, 1))
# (2)
  set.seed(35)
  xx1 <- rnorm(10, mean = 2, sd = 1)
  xx2 <- rnorm(10, mean = 3, sd = 1)
  xx3 <- rnorm(10, mean = 4, sd = 2)
  data1 <- data.frame(x1 = xx1, x2 = xx2, x3 = xx3)
# (3)
  stars(data1, locations = c(0, 0), key.loc = c(0, 0), lwd = 1)
}
```

(1) The graphics area is set.
(2) Simulation data (xx1, xx2, and xx3) are constructed. They are put together in the data frame data1.

(3) The command **stars()** illustrates a radar chart. The argument **locations = c(0, 0)** sets the position of the graph. Since all data are drawn at the center of the area in this example, **c(0, 0)** is set. The argument **key.loc = c(0, 0)** specifies the locations of the outlines. Since all the outlines are drawn at the center of the area in this example, **c(0, 0)** is set.

A parallel chart is also a means of depicting the relationship between data with plural variables. Fig. 2.13 (right) given by Program (2 - 25) is an example.

```
Program (2 - 25)
function() {
# (1)
  par(mai = c(1, 1, 1, 1), omi = c(0, 0, 0, 0))
# (2)
  set.seed(35)
  xx1 <- rnorm(10, mean = 2, sd = 1)
  xx2 <- rnorm(10, mean = 3, sd = 1)
  xx3 <- rnorm(10, mean = 4, sd = 2)
  data1 <- data.frame(x1 = xx1, x2 = xx2, x3 = xx3)
# (3)
  xmin1 <- min(xx1, xx2, xx3) - 0.5
  xmax1 <- max(xx1, xx2, xx3) + 0.5
# (4)
  plot(c(1, 2, 3), c(0, 0, 0), xlab = "", ylab = "", type = "n",
    axes = F, xlim = c(0.6, 3.4), ylim = c(xmin1, xmax1))
# (5)
  for(ii in 1:10){
    lines(c(1, 2, 3), c(xx1[ii], xx2[ii], xx3[ii]))
    points(c(1, 2, 3), c(xx1[ii], xx2[ii], xx3[ii]),
      pch = LETTERS[ii])
  }
# (6)
  axis(2)
  box(lwd = 1)
# (7)
  par(new = T)
# (8)
  name1 <- c("x1", "x2", "x3")
# (9)
  barplot(c(0, 0, 0), xlab = "", ylab = "",  names.arg = name1,
    axes = F, ylim = c(xmin1, xmax1), border = "white")
}
```

(1) The graphics area is set.
(2) Simulation data (xx1, xx2, and xx3) are constructed. They are put together in the data frame **data1**.

(3) The minimum value of the vertical axis of the graph is set as **xmin1**; its maximum value is set as **xmax1**.

(4) The coordinate axes are set.

(5) The values of **xx1**, **xx2**, and **xx3** are drawn in the form of a parallel chart.

(6) The command **axis(2)** draws the y-axis. The command **box()** draws lines surrounding the graph.

(7) The command **par(new = T)** indicates that the current graph will remain as it is and a new graph will be superimposed.

(8) **xx1**, **xx2**, and **xx3** are named **name1**.

(9) The command **barplot()** adds the explanation of the horizontal axis of the graph. The argument **names.arg = name1** indicates that the explanation of the horizontal axis is **name1**. Since the first argument is **c(0, 0, 0)**, the height of the three bars of the bar plot is 0. This means that the three bars are three horizontal lines. However, **border = "white"** makes the three horizontal lines invisible. Hence, **barplot()** just adds the explanations of the horizontal axis.

2.10   FUNCTIONS OF ONE VARIABLE

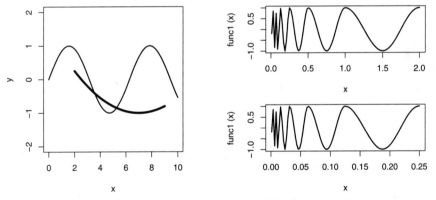

Figure 2.14   Function given by Program (2 - 26) (left). Self-similar function given by Program (2 - 27) (right).

To draw the functional relationship of a function with one variable in a graph, Program (2 - 26), for example, is used. Fig. 2.14 (left) is the result of this R program.

Program (2 - 26)
```
function() {
# (1)
  par(mai = c(1, 1, 1, 1), omi = c(0, 0, 0, 0))
```

```
# (2)
  curve(sin, 0, 10, lwd = 2, xlab = "x", ylab = "y",
   xlim = c(0, 10), ylim = c(-2, 2))
# (3)
  func1 <- function(x){
    yy <- 0.05 * (x - 7)^2 - 1
  return(yy)
  }
# (4)
   curve(func1, 2, 9, add = T, lwd = 4)
}
```

(1) The graphics area is set.
(2) The command `curve()` draws the values calculated using `sin()`. The range of the variable is between 0 and 10.
(3) The function `func1()` is defined.
(4) The values calculated using `func1()` are drawn. The range of the variable is between 2 and 9.

An R function is defined in a recursive manner. This feature is taken advantage of in Fig. 2.14 (right) given by Program (2 - 27).

```
Program (2 - 27)
function() {
# (1)
  par(mfrow = c(2, 1), mai = c(1, 1, 0.2, 0.2),
   omi = c(0, 0, 0, 0))
# (2)
  func1 <- function(qq){
# (3)
     func2 <- function(tt){
       if(tt >= 1){
         uu <- cos(2 * pi * tt)
         return(uu)
       }
       else {
         tt <- tt * 2
         uu <- func2(tt)
         return(uu)
       }
     }
# (4)
     nd <- length(qq)
     ss <- NULL
     for(ii in 1:nd){
       ss[ii] <- func2(qq[ii])
     }
```

```
    return(ss)
  }
# (5)
  curve(func1, 0.01, 2, lwd = 2)
  curve(func1, 0.00125, 0.25, lwd = 2)
}
```

(1) The graphics area is set.

(2) The function `func1()` is defined. The command `curve()` gives a vector consisting of plural data as a variable for calculating the values of the function and receives the results of the calculation as a vector. Hence, if a function that does not allow a variable in the vector format is used, each element of the vector is extracted to calculate the result of the function. This is the reason for the call for `func2()` in `func1()`.

(3) The function `func2()` is defined. If the value of the argument (a scalar not a vector) is equal to or larger than 1, the value calculated using `cos()` is output. If it is less than 1, the value of the argument is doubled and the function `func2()` is called in a recursive manner.

(4) Each element of `qq` (a vector), which is an argument of `func1()`, is extracted, and `func2()` carries out a calculation using the value.

(5) The command `curve()` draws the values yielded using `func1()` with the argument ranging from 0.01 to 2. Furthermore, the values calculated using `func1()` using the argument ranging from 0.00125 to 0.25 are depicted in the graph. These two graphs look the same except for the values along the x-axis. This shows that this function is self-similar.

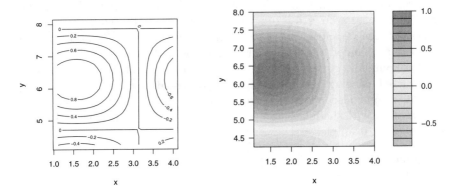

Figure 2.15   Contour given by Program (2 - 28) (left). Contour given by Program (2 - 29) (right).

2.11 FUNCTIONS OF TWO VARIABLES

An R command draws a contour. Fig. 2.15 (left) given by Program (2 - 28) exemplifies it.

```
Program (2 - 28)
function() {
# (1)
  par(mai = c(1, 1, 1, 1), omi = c(0, 0, 0, 0))
# (2)
  fun1 <- function(x, y){
    zz <- sin(x) * cos(y)
  return(zz)
  }
# (3)
  nx <- 30
  ny <- 16
  xx <- seq(from = 1.1, to = 4, length = nx)
  yy <- seq(from = 4.25, to = 8, length = ny)
# (4)
  zz <- outer(xx, yy, fun1)
# (5)
  contour(xx, yy, zz, xlab = "x", ylab = "y")
}
```

(1) The graphics area is set.
(2) The function fun1() is defined.
(3) The points of the x-axis for deriving the values of func1() and for illustrating a contour are given as xx. Those of the y-axis are given as yy. The function seq(from = 1.1, to = 4, length = nx) indicates the equally spaced nx values between 1.1 and 4. Hence, the elements of xx are $(1.1, 1.2, 1.3, , \ldots, 4.0)$. In a similar manner, those of yy are $(4.25, 4.50, 4.75, \ldots, 8.00)$.
(4) The command outer() calculates the values of func1() at the grid points constructed using xx and yy; the results are stored as zz.
(5) The command contour() illustrates a contour using zz.

A contour with shading is also possible. Fig. 2.15 (right) given by Program (2 - 29) is an example.

```
Program (2 - 29)
function() {
# (1)
  par(mai = c(1, 1, 1, 1), omi = c(0, 0, 0, 0))
# (2)
  fun1 <- function(x, y){
    zz <- sin(x) * cos(y)
  return(zz)
```

```
    }
# (3)
  nx <- 30
  ny <- 16
  xx <- seq(from = 1.1, to = 4, length = nx)
  yy <- seq(from = 4.25, to = 8, length = ny)
# (4)
  zz <- outer(xx, yy, fun1)
# (5)
  filled.contour(xx, yy, zz, xlab = "x", ylab = "y")
}
```

(1) The graphics area is set.
(2) The function **fun1()** is defined.
(3) The points of the x-axis for deriving the values of **func1()** and illustrating a contour are given as **xx**. Those of the y-axis are given as **yy**.
(4) The command **outer()** calculates the values of **func1()** at the grid points constructed using **xx** and **yy**; the results are stored as **zz**.
(5) The command **filled.contour()** illustrates a contour with shading using **zz**.

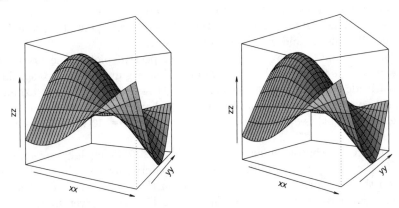

Figure 2.16 Perspective plot given by Program (2 - 30) (left). Perspective plot from a slightly different viewpoint (right).

An example of a perspective plot is Fig. 2.16 (left) (right). When Fig. 2.16 (left) is viewed with the left eye and Fig. 2.16 (right) is viewed with the right eye, the object is viewed stereoscopically. Fig. 2.16 (left) is illustrated using Program (2 - 30).

```
Program (2 - 30)
function() {
# (1)
  par(mai = c(1, 1, 1, 1), omi = c(0, 0, 0, 0))
```

```
# (2)
  fun1 <- function(x, y){
    zz <- sin(x) * cos(y)
  return(zz)
  }
# (3)
  nx <- 30
  ny <- 16
  xx <- seq(from = 1.1, to = 4, length = nx)
  yy <- seq(from = 4.25, to = 8, length = ny)
# (4)
  zz <- outer(xx, yy, fun1)
# (5)
  persp(xx, yy, zz, theta = 30, phi = 0, ltheta = 280,
    shade = 0.3, col = "yellow" )
}
```

(1) The graphics area is set.
(2) The function `fun1()` for giving the form of a perspective plot is defined.
(3) The points of the x-axis for deriving the values of `func1()` and illustrating a contour are given as `xx`. Those of the y-axis are given as `yy`.
(4) The command `outer()` calculates the values of `func1()` at the grid points constructed using `xx` and `yy`; the results are stored as `zz`.
(5) The command `persp()` illustrates a perspective plot. The argument `theta = 30` sets the azimuth of a viewpoint. The argument `phi = 0` specifies the colatitude of a viewpoint. The argument `ltheta = 280` sets the azimuth of a light source. The argument `shade = 0.3` shows the thickness of a shade.

Fig. 2.16 (right) is obtained by substituting `theta = 30` with `theta = 35` in `persp()` in (6).

A color-coded contour is also possible. Fig. 2.17 (left)(right) exemplifies this. In Fig. 2.17 (left), a different color is painted for each interval of the scale on the contour. In Fig. 2.17 (right), on the other hand, continuously varying color is painted on the contour. Fig. 2.17 (left) is given by Program (2 - 31).

```
Program (2 - 31)
function() {
# (1)
  par(mai = c(1, 1, 1, 1), omi = c(0, 0, 0, 0))
# (2)
  fun1 <- function(x, y){
    zz <- sin(x) * cos(y)
  return(zz)
  }
# (3)
  colfunc1 <- function(x) {
```

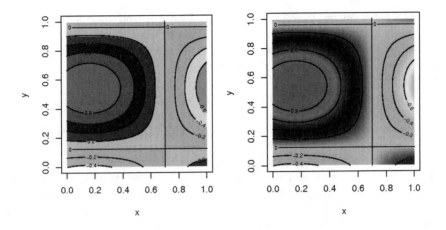

Figure 2.17 Contour given by Program (2 - 31) (left). Contour given by Program (2 - 32) (right).

```
    xave <- (x[-1, -1] + x[-1, -(ncol(x) - 1)] + x[-(nrow(x) -1),
     -1] + x[-(nrow(x) -1), -(ncol(x) - 1)]) / 4
    col1 <- rainbow(10)
    colors <-  col1[cut(xave, breaks = seq(from = -1, to = 1,
     by = 0.2), include.lowest = T)]
    return(colors)
  }
# (4)
  nx <- 1000
  ny <- 1000
  xx <- seq(from = 1.1, to = 4, length = nx)
  yy <- seq(from = 4.25, to = 8, length = ny)
# (5)
  zz <- outer(xx, yy, fun1)
# (6)
  persp(zz, theta = 0, phi = 90, lphi = 90, r = 10000,
   zlim = c(0,1000), col = colfunc1(zz), border = NA, box = F)
# (7)
  par(new = T)
# (8)
  contour(zz, xlab = "x", ylab = "y", levels =
   seq(from = -1, to = 1, by = 0.2))
}
```

(1) The graphics area is set.
(2) The function fun1() is defined.

(3) The function `colfunc1()` is defined. This function specifies the painting colors for respective heights of the contour when a perspective plot is drawn using `persp()`. The colors given by dividing the colors of `rainbow()` into 10 steps are assigned to the respective regions given by the separation $\{-1, -0.8, -0.6, \ldots, 1\}$.

(4) The points of the x-axis for deriving the values of `func1()` and illustrating a contour are given as `xx`. Those of the y-axis are given as `yy`.

(5) The command `outer()` calculates the values of `func1()` at the grid points constructed using `xx` and `yy`.

(6) The command `persp()` illustrates a perspective plot. The argument `theta = 0` indicates that the perspective plot is viewed from the front. The argument `phi = 90` indicates that the perspective plot is viewed directly from above. The argument `lphi = 90` means that the light is shed directly from above. The argument `r = 10000` means that the perspective plot is drawn substantially with no perspective.

(7) The command `par(new = T)` describes that the current graph will remain as it is and a new graph will be superimposed.

(8) The command `contour()` illustrates a contour using `zz`. The argument `levels = seq(from = -1, to = 1, by = 0.1)` indicates the intervals of the contour.

Fig. 2.17 (right) is obtained by replacing the parts in which `col1` and `colors` are given in (3) of Program (2 - 30) with the following program.

```
col1 <- rainbow(200)
colors <-  col1[cut(xave, breaks = seq(from = -1, to = 1,
  by = 0.01), include.lowest = T)]
```

A contour can be drawn on the curved surface of a perspective plot. Fig. 2.18 (left) given by Program (2 - 32) is an example.

Program (2 - 32)
```
function ()
{
# (1)
  fun1 <- function(x, y){
    zz <- sin(x) * cos(y)
  return(zz)
  }
# (2)
  colfunc1 <- function(x) {
  xave <- (x[-1, -1] + x[-1, -(ncol(x) - 1)] + x[-(nrow(x) -1),
    -1] + x[-(nrow(x) -1), -(ncol(x) - 1)]) / 4
  br1 <- seq(from = -0.8, to = 1, by=0.2)
  col1 <- cm.colors(length(br1)-1)
  colors <- col1[cut(xave, breaks = br1, include.lowest = T)]
   return(colors)
```

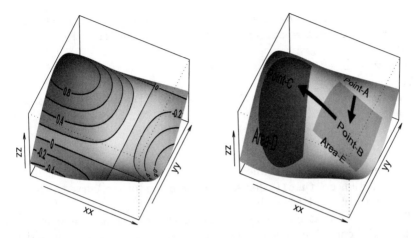

Figure 2.18   Perspective plot in which an image of a contour is attached to the curved surface, given by Program (2 - 32) (left). Perspective plot in which a black-and-white image is colored and attached to the curved surface, given by Program (2 - 33) (right).

```
}
# (3)
  colfunc2 <- function(x) {
  xave <- (x[-1, -1] + x[-1, -(ncol(x) - 1)] + x[-(nrow(x) -1),
   -1] + x[-(nrow(x) -1), -(ncol(x) - 1)]) / 4
  colors <- rep("transparent", length=length(xave))
  colors[0 <= xave &  200 > xave] <- "black"
  return(colors)
}
# (4)
  par(mai = c(1, 1, 1, 1), omi = c(0, 0, 0, 0))
# (5)
  map1 <- read.csv(file = "d:\\GraphicsR\\map1.txt", header=F)
# (6)
  map1 <- matrix(unlist(map1), nrow = 683)
  map1 <- map1[ , seq(from = 516, to = 1, by = -1)]
# (7)
  nx <- dim(map1)[1]
  ny <- dim(map1)[2]
  xx <- seq(from = 1.1, to = 4, length = nx)
  yy <- seq(from = 4.25, to = 8, length = ny)
# (8)
  zz <- outer(xx, yy, fun1)
# (9)
```

```
persp(xx, yy, zz, theta = 25, phi = 60, ltheta = 70, lphi=20,
  shade = 0.0, col = colfunc1(zz),r = 100, border = F)
# (10)
par(new = T)
# (11)
persp(xx, yy, zz, theta = 25, phi = 60, ltheta = 70, lphi=20,
  shade = 0.0, col = colfunc2(map1),r = 100, border = F,
  axes = F, box = F)
}
```

(1) The function `fun1()` is defined.
(2) The function `colfunc1()` for setting the colors of the curved surface of the first perspective plot is defined.
(3) The function `colfunc2()` for setting black for a contour and numbers drawn on the curved surface of the superimposed perspective plot and the transparency for the remaining area on the same surface is defined.
(4) The graphics area is set.
(5) The text file map1.txt for drawing a contour and numbers on the curved surface of the superimposed perspective plot is retrieved and named `map1`.
(6) Since `map1` has a vector format, it is converted to a matrix format. Furthermore, the image is inverted.
(7) The coordinates of the grid points for calculating the value of the function `fun1()` are derived. The number of grid points is identical to `map1`.
(8) The command `outer()` calculates the values of `func1()` at the grid points; the results are stored as `zz`.
(9) The command `persp()` illustrates a perspective plot.
(10) The command `par(new = T)` describes that the current graph will remain as it is and a new graph will be superimposed.
(11) A perspective plot is superimposed. The contour and numbers are drawn in black in this perspective plot and the remaining parts are transparent.

The command `contour()` draws a contour to illustrate Fig. 2.17 (left)(page 103). On the other hand, an image drawn using `contour()` is captured on the display of a personal computer and the resultant image is stored as `map1.txt` to be attached to the curved surface in a perspective plot.

The technique adopted in Program (2 - 32) allows users to put a map-like image on the curved surface in a perspective plot. Fig. 2.18 (right) given by Program (2 - 33) is a simple example.

Program (2 - 33)
```
function ()
{
# (1)
  fun1 <- function(x, y){
    zz <- sin(x) * cos(y)
  return(zz)
  }
```

```
# (2)
  colfunc3 <- function(x) {
  xave <- (x[-1, -1] + x[-1, -(ncol(x) - 1)] + x[-(nrow(x) -1),
   -1] + x[-(nrow(x) -1), -(ncol(x) - 1)]) / 4
  col1 <- heat.colors(max(x) + 1)
  colors <- col1[cut(xave, breaks = length(col1),
   include.lowest = T)]
  colors[0 <= xave &  100 > xave] <- "black"
  colors[100 <= xave &  150 > xave] <- "red"
  colors[150 <= xave &  200 > xave] <- "blue"
  colors[200 <= xave &  240 >= xave] <- "green"
  colors[240 <= xave &  255 >= xave] <- "yellow"
  return(colors)
}
# (2)
  par(mai = c(1, 1, 1, 1), omi = c(0, 0, 0, 0))
# (3)
  map2 <- read.csv(file = "d:\\GraphicsR\\map2.txt",
   header = F)
# (4)
  map2 <- matrix(unlist(map2), nrow = 582)
  map2 <- map2[ , seq(from = 435, to = 1, by = -1)]
# (5)
  nx <- dim(map2)[1]
  ny <- dim(map2)[2]
  xx <- seq(from = 1.1, to = 4, length = nx)
  yy <- seq(from = 4.25, to = 8, length = ny)
# (6)
  zz <- outer(xx, yy, fun1)
# (7)
  persp(xx, yy, zz, theta = 25, phi = 60, ltheta = 70, lphi=20,
   shade = 0.0, col = colfunc3(map2),r = 100, border = F)
}
```

(1) The function **fun1**() is defined.
(2) The function **colfunc3**() for specifying the colors of the curved surface of a perspective plot is defined. This function allows users to put a colored image on the curved surface in a perspective plot using a black-and-white image (map2.txt).
(3) The text file map2.txt for producing an image painted on the curved surface of a perspective plot is retrieved and named **map2**. The image file is a grayscale image.
(4) Since **map2** has a vector format, it is converted to a matrix format. Furthermore, it is inverted.

(5) The coordinates of grid points for calculating the values of the function fun1() are derived. The number of grid points is coincident with that of elements of map2.

(6) The command outer() calculates the values of func1() at the grid points constructed using xx and yy; the results are stored as zz.

(7) The command persp() illustrates a perspective plot.

2.12   MAP GRAPHS

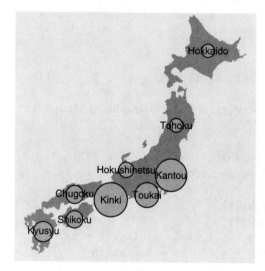

Figure 2.19   Map of Japan with region-by-region population densities, given by Program (2 - 34).

The command image() is also useful for drawing a map. Graphic symbols and letters can be added on the map. Fig. 2.19 given by Program (2 - 34) exemplifies this. This map shows the region-by-region population densities of Japan on September 30, 1998.

```
Program (2 - 34)
function() {
# (1)
  par(mai = c(0.3, 0.3, 0.3, 0.3), omi = c(0, 0, 0, 0), cex = 0.7)
# (2)
  japan1 <- read.csv(file = "d:\\GraphicsR\\japan1.txt", header=F)
# (3)
  japan1 <- unlist(japan1)
  japan1 <- matrix(japan1, nrow = 1000)
```

```
  japan1 <- japan1[ , seq(from = 1010, to = 1, by = -1)]
# (4)
  japan1 <- japan1[251:900, 281:990]
  image(japan1, col = c("yellow", "skyblue"), axes = F)
  print(par("usr"))
# (5)
  hokkaido <- sqrt(318) * 0.0015
  touhoku <- sqrt(342) * 0.0015
  hokushinetsu <- sqrt(431) * 0.0015
  kantou <- sqrt(1902) * 0.0015
  toukai <- sqrt(1111) * 0.0015
  kinki <- sqrt(2101) * 0.0015
  chugoku <- sqrt(598) * 0.0015
  shikoku <- sqrt(552) * 0.0015
  kyushyu <- sqrt(751) * 0.0015
# (6)
  symbols(0.8, 0.85, circles = hokkaido, add = T, bg = "pink",
   inches = F)
  text(0.8, 0.85, "Hokkaido")
  symbols(0.67, 0.55, circles = touhoku, add = T, bg = "pink",
   inches = F)
  text(0.67, 0.55, "Tohoku")
  symbols(0.463, 0.37, circles = hokushinetsu, add = T, bg = "pink",
   inches = F)
  text(0.463, 0.37, "Hokushinetsu")
  symbols(0.65, 0.35, circles = kantou, add = T, bg = "pink",
   inches = F)
  text(0.65, 0.35,  "Kantou")
  symbols(0.55, 0.27, circles = toukai, add = T, bg = "pink",
   inches = F)
  text(0.55, 0.27, "Toukai")
  symbols(0.4, 0.25, circles = kinki, add = T, bg = "pink",
   inches = F)
  text(0.4, 0.25, "Kinki")
  symbols(0.25, 0.27, circles = chugoku, add = T, bg = "pink",
   inches = F)
  text(0.25, 0.27, "Chugoku")
  symbols(0.25, 0.17, circles = shikoku, add = T, bg = "pink",
   inches = F)
  text(0.25, 0.17, "Shikoku")
  symbols(0.12, 0.12, circles = kyushyu, add = T, bg = "pink",
   inches = F)
  text(0.12, 0.12, "Kyusyu")
}
```

(1) The graphics area is set. The argument `cex = 0.7` sets the size of letters.
(2) The data for drawing the map of Japan is retrieved and named japan1.
The file japan1.txt is text data consisting of the values of 0 or 255. For the
origin of japan1.txt, refer to the Appendix.
(3) After `japan1` is converted to numerical values, the result is transformed
into a matrix format. A matrix (1000 × 1010) is obtained. Furthermore,
the values of respective rows are inverted. This is because `image()` displays
an inverse image compared with the ordinary image display on a personal
computer.
(4) Peripheral parts of the map of Japan are omitted. Then, `image()` displays
the map of Japan. The argument `col = c("yellow", "skyblue")` paints the
sea yellow and paints the land sky-blue. The command `print(par("usr"))`
outputs the maximum and minimum of the coordinate axes set by `image()`.
In this example, this command outputs the following:

```
-0.0007704160  1.0007704160 -0.0007052186  1.0007052186
```

That is, the ranges of both x-axis and y-axis are roughly between 0 and 1.
Graphic symbols and letters can be added on the basis of this result.
(5) Region-by-region population densities are given. For example, the regional
population density of the Hokkaido area is 318. Since the areas of circles stand
for regional population densities, their square roots are derived. The value of
`* 0.0015` is a multiplier factor for making the sizes of circles appropriate.
For the origin of the regional population density data, refer to the Appendix.
(6) The command `symbols()` shows the the regional population densities as
areas of the circles. The argument `bg = "pink"` paints the circles in pink.
The argument `inches = F` indicates that the radius of a circle (`circles =`)
is specified using the scale of the x-axis. The argument `text()` adds letters.

The command `persp()` realizes a bird's-eye-view map. Users can add bar
plots and letters in the graph. Fig. 2.20 exemplifies this. When Fig. 2.20
(left) is viewed with the left eye and Fig. 2.20 (right) is viewed with the right
eye, the object is viewed stereoscopically. Fig. 2.20 (left) is yielded using
Program (2 - 35).

```
Program (2 - 35)
function() {
# (1)
  par(mai = c(0.3, 0.3, 0.3, 0.3), omi = c(0, 0, 0, 0))
# (2)
  japan1 <- read.csv(file = "d:\GraphicsR\\japan1.txt", header=F)
# (3)
  colfunc1 <- function(x) {
    xave <- (x[-1, -1] + x[-1, -(ncol(x) - 1)] + x[-(nrow(x) -1),
      -1] + x[-(nrow(x) -1), -(ncol(x) - 1)]) / 4
    colors <- rep("skyblue", length(xave))
    colors[xave == 0] <- "white"
```

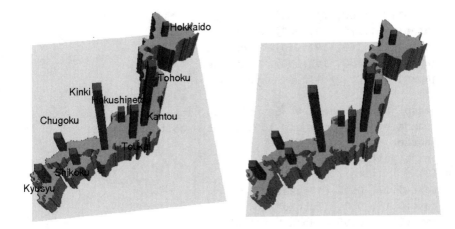

Figure 2.20   Map of Japan with region-by-region population densities, given by Program (2 - 35) (left). Graph the same as the left one except for a slightly different viewpoint (right).

```
    colors[xave == 255] <- "green"
    colors[xave > 255] <- "red"
    return(colors)
  }
# (4)
  japan1 <- unlist(japan1)
  japan1 <- matrix(japan1, nrow = 1000)
  japan1 <- japan1[,seq(from = 1010, to = 1, by = -1)]
  japan1 <- japan1[c(251:900), 281:990]
# (5)
  hokkaido <- 318 / 2101 * 700 + 255
  touhoku <- 342 / 2101 * 700 + 255
  hokushinetsu <- 431 / 2101 * 700 + 255
  kantou <- 1902 / 2101 * 700 + 255
  toukai <- 1111 / 2101 * 700 + 255
  kinki <- 2101 / 2101 * 700 + 255
  chugoku <- 598 / 2101 * 700 + 255
  shikoku <- 552 / 2101 * 700 + 255
  kyushyu <- 751 / 2101 * 700 + 255
# (6)
  japan1[491:510, 591:610] <- hokkaido
  japan1[431:450, 391:410] <- touhoku
  japan1[311:330, 271:290] <- hokushinetsu
  japan1[381:400, 251:270] <- kantou
```

```
  japan1[341:360, 211:230] <- toukai
  japan1[251:270, 181:200] <- kinki
  japan1[121:140, 191:210] <- chugoku
  japan1[161:180, 141:160] <- shikoku
  japan1[71:90, 91:110] <- kyushyu
# (7)
  persp(japan1, theta = -5, phi = 65, ltheta = 40, lphi = 60,
    shade = -5, zlim = c(0,1000), col = colfunc1(japan1),
    border = NA, box = F)
  print(par("usr"))
# (8)
  text(0.17, 0.12, "Hokkaido")
  text(0.13, 0.0, "Tohoku")
  text(0.0, -0.05, "Hokushinetsu")
  text(0.1, -0.09, "Kantou")
  text(0.036, -0.16, "Toukai")
  text(-0.1, -0.03, "Kinki")
  text(-0.15, -0.1, "Chugoku")
  text(-0.12, -0.22, "Shikoku")
  text(-0.2, -0.26, "Kyusyu")
}
```

(1) The graphics area is set.

(2) The data (japan1.txt) for producing the map of Japan is retrieved and named `japan1`.

(3) The function `colfunc1()` is defined. This function specifies the colors depending on the height when `persp()` illustrates a perspective plot. When the values of the elements of `japan1` (heights) are 0 (the sea), `"white"` is applied. When the values of the elements of `japan1` are 255 (land), `"green"` is applied. When the values of the elements of `japan1` are 255 (a bar plot), `"red"` is specified. Otherwise (pillars supporting the land), `"skyblue"` is applied.

(4) After `japan1` is converted to numerical values, the result is transformed into a matrix format. A matrix (1000 × 1010) is obtained. Furthermore, the values of respective rows are inverted.

(5) Region-by-region population densities are given. For example, the regional population density of Hokkaido area is `318`. The calculation `/ 2101 * 700 + 255` makes the heights of the bars of the bar plots appropriate, Since the heights of the land are 255, "+ 255" makes the heights of the bars of the bar plots 0, when the values for the bar plots are 0.

(6) The region-by-region population densities are written in the neighborhood of the bottom of the bars of the bar plots.

(7) Peripheral parts of the map of Japan are omitted. The command `persp()` illustrates a bird's-eye-view map and bar plots. The command `print(par`

("usr")) outputs the maximum and minimum of the coordinate axes set by
image() . In this example, this command outputs the following:

-0.4170055 0.4768157 -0.4001116 0.4918795

That is, the ranges of both x-axis and y-axis are roughly between −0.41 and
0.48.
(8) The argument text() adds letters.
   When (8) is removed from this R program and the line containing persp()
in (7) is substituted with the following command, Fig. 2.20 (right) is obtained.

```
persp(japan1, theta = 0, phi = 65, ltheta = 40, lphi = 60,
  shade = -5, zlim = c(0,1000), col = colfunc1(japan1),
  border = NA, box = F)
```

2.13 HISTOGRAMS OF TWO VARIABLES

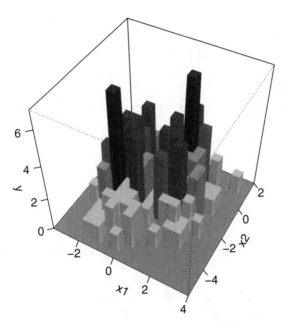

Figure 2.21   Histogram of two variables given by Program (2 - 36).

   The perspective plot can be applied to drawing a histogram with two variables. For example, Fig. 2.21 given by Program (2 - 36) realizes this.

Program (2 - 36)

```r
function ()
{
# (1)
  colfunc4 <- function(x) {
  xave <- (x[-1, -1] + x[-1, -(ncol(x) - 1)] +
   x[-(nrow(x) -1), -1] + x[-(nrow(x) -1), -(ncol(x) - 1)]) / 4
  col1 <- rev(heat.colors(max(x) + 2))
  colors <- col1[cut(xave, breaks = length(col1),
   include.lowest = T)]
  n1 <- dim(x)[1]
  n2 <- dim(x)[2]
  for(ii in 1:(n1 - 1)){
    for(jj in 1:(n2 - 1)){
      v1 <- sd(c(x[ii, jj], x[ii, jj + 1] ))
      if(x[ii,jj] != x[ii, jj + 1]){
         colors[-n1 + 1 + jj * (n1 - 1)+ii] <- col1[x[ii, jj + 1]
          + 1]
      }
    }
  }
  for(ii in 1:(n1 - 1)){
    for(jj in 1:(n2 - 1)){
      v1 <- sd(c(x[ii, jj], x[ii + 1, jj] ))
      if(v1 != 0){
         colors[-n1 + 1 + jj * (n1 - 1) + ii] <- col1[x[ii, jj + 1]
          + 1]
      }
    }
  }
  colors[xave == 0] <- "gray"
  return(colors)
}
# (2)
 par( mai = c(0.3, 0.3, 0.1, 0.1), omi = c(0, 0, 0, 0))
# (3)
    set.seed(703)
    xx1 <- rnorm(100, mean = -0.8, sd = 0.7)
    xx2 <- rnorm(100, mean = -2.1, sd = 1.2)
    xx1 <- c(xx1 ,rnorm(100, mean = 1.3, sd = 0.7))
    xx2 <- c(xx2, rnorm(100, mean = -1.1, sd = 1.2))
# (4)
    br1 <- pretty(n = 10, xx1)
    dx1 <- br1[2] -  br1[1]
    br1 <- c(min(br1) - dx1, br1, max(br1) + dx1)
    br2 <- pretty(n = 10, xx2)
```

```
      dx2 <- br2[2] -  br2[1]
      br2 <- c(min(br2) - dx2, br2, max(br2) + dx2)
# (5)
    his2 <- matrix(rep(0, length = (length(br1) - 1) *
    (length(br2) - 1)), ncol = length(br2) - 1)
    for(kk in 1:length(xx1)){
      for(jj in 1:(length(br2) - 1)){
        for(ii in 1:(length(br1) - 1)){
          if((br1[ii] <= xx1[kk]) & (br1[ii + 1] > xx1[kk]) &
          (br2[jj] <= xx2[kk]) & (br2[jj+1] > xx2[kk]))
          {
            his2[ii,jj] <- his2[ii,jj] + 1
          }
        }
      }
    }
# (6)
    finep1 <- seq(from = min(br1), to = max(br1), length =
    dim(his2)[1] * 10)
    finep2 <- seq(from = min(br2), to = max(br2), length =
    dim(his2)[2] * 10)
    fineh2d <- matrix(rep(-20, length = length(finep1) *
    length(finep2)), ncol = length(finep2))
    for(ii in 1:length(finep1)){
      for(jj in 1:length(finep2)){
        fineh2d[ii,jj] <- his2[(ii+9) %/% 10,(jj+9) %/% 10]
      }
    }
# (7)
    persp(finep1, finep2, fineh2d, theta = 30, phi = 45,
    ltheta = 50, lphi=50, shade=0, r = 5, col =
    colfunc4(fineh2d), ticktype = "detailed", xlab = "x1",
    ylab = "x2", zlab = "y", border = F)
}
```

(1) The function `colfunc4()` for specifying the colors of bars of a histogram is defined. The colors of lateral aspects of the bars are set to be the same as those of the top aspects of the bars.
(2) The graphics area is set.
(3) Simulation data (`xx1` and `xx2`) are produced.
(4) The positions of break points for the two variables are set as `br1` and `br2`.
(5) The frequency of the histogram of the two variables is derived and named `his2`.

(6) To illustrate a histogram of two variables using a perspective plot given by `persp()`, the values for the top of the respective bars should be constant. This consideration yields `finep1`, `finep2`, and `fineh2d`.

(7) A perspective plot given by `persp()` results in a histogram of two variables.

2.14 TIME SERIES GRAPHS OF TWO VARIABLES

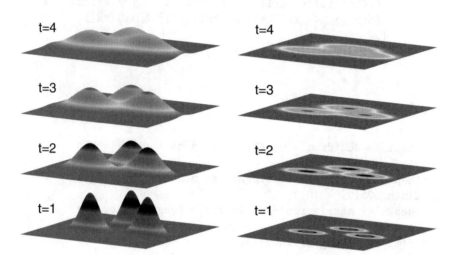

Figure 2.22 Perspective plot for showing time series of distributions of heat, given by Program (2 - 37) (left). Graph yielded by shrinking the z-axis of the left graph (right).

A perspective plot allows users to look across the sweep of a time series of a function with two variables. Fig. 2.22 shows a time series of solutions of the heat-transfer function on an infinite plane. Fig. 2.22 (left) is produced using Program (2 - 37).

```
Program (2 - 37)
function() {
# (1)
  par(omi = c(0, 0, 0, 0))
# (2)
  heatf1 <- function(tt, ex1, ey1, xd, yd){
    nd <- length(xd)
    zz1 <- 0
    for (ii in 1:nd){
      zz1 <- zz1 + exp( - ((xd[ii] - ex1)^2 + (yd[ii] -
      ey1)^2) / tt) / sqrt(tt)
```

```
    }
  return(zz1)
}
# (3)
  colfunc1 <- function(x){
    xave <- (x[-1, -1] + x[-1, -(ncol(x) - 1)] + x[-(nrow(x)
     -1), -1] + x[-(nrow(x) -1), -(ncol(x) - 1)]) / 4
    col1 <- rainbow(500)
    colors <- col1[cut(xave, breaks = seq(from = 0, to = 1,
     length = 501), include.lowest = T)]
    return(colors)
  }
# (4)
   xd <- NULL
   yd <- NULL
   xd[1] <- 3.2
   yd[1] <- 8.3
   xd[2] <- 5.1
   yd[2] <- 13.0
   xd[3] <- 7.3
   yd[3] <- 9.8
# (5)
   nx <- 200
   ny <- 400
   ex <- seq(from = 0.05, to = 10, length = nx)
   ey <- seq(from = 0.05, to = 20, length = ny)
# (6)
  par(mai = c(0, 0, 3, 0))
  tt <- 1
  zz <- matrix(rep(0, nx * ny), ncol = ny)
  for(ii in 1:nx){
    for(jj in 1:ny){
      zz[ii,jj] <- heatf1(tt, ex[ii], ey[jj], xd, yd)
    }
  }
  persp(zz, theta = 30, phi = 10, ltheta = 0, lphi = 90,
   shade = -5, r = 1000, zlim = c(0,4), col = colfunc1(zz),
   border = NA, box = F)
  text(-0.001, -0.0007, "t=1")
# (7)
  par(new = T)
  par(mai = c(1 , 0, 2, 0))
  tt <- 2
  zz <- matrix(rep(0, nx * ny), ncol = ny)
  for(ii in 1:nx){
```

```
  for(jj in 1:ny){
    zz[ii,jj] <- heatf1(tt, ex[ii], ey[jj], xd, yd)
  }
}
persp(zz, theta = 30, phi = 10, ltheta = 0, lphi = 90,
  shade = -5, r = 1000, zlim = c(0, 4), col = colfunc1(zz),
  border = NA, box = F)
text(-0.001, -0.0007, "t=2")
# (8)
par(new = T)
par(mai = c(2, 0, 1, 0))
tt <- 3
zz <- matrix(rep(0, nx * ny), ncol = ny)
for(ii in 1:nx){
  for(jj in 1:ny){
    zz[ii,jj] <- heatf1(tt, ex[ii], ey[jj], xd, yd)
  }
}
persp(zz, theta = 30, phi = 10, ltheta = 0, lphi=90,
  shade = -5, r = 1000, zlim = c(0, 4), col = colfunc1(zz),
  border = NA, box = F)
text(-0.001, -0.0007, "t=3")
# (9)
par(new = T)
par(mai = c(3, 0, 0, 0))
tt <- 4
zz <- matrix(rep(0, nx * ny), ncol = ny)
for(ii in 1:nx){
  for(jj in 1:ny){
      zz[ii,jj] <- heatf1(tt, ex[ii], ey[jj], xd, yd)
  }
}
persp(zz, theta = 30, phi = 10, ltheta = 0, lphi = 90,
  shade = -5, r = 1000, zlim = c(0, 4), col = colfunc1(zz),
  border = NA, box = F)
text(-0.001, -0.0007, "t=4")
}
```

(1) The graphics area is set.
(2) The function heatf1() for solving the heat-transfer function on an infinite plane is defined. xd (coordinates of x-axis) and yd (coordinates of x-axis) indicate the position of delta functions when the distribution of heat on tt(time) = 0 is represented by a set of the delta functions. The values of ex1 (a scalar) and ey1 (a scalar), respectively, indicate the coordinates of the x-axis and y-axis at a position where the amount of heat is calculated. If a positive value

is given to tt, the amount of heat at the position given by ex1 and ey1 is derived.

(3) The function colfunc1() for setting different colors according to the height in a perspective plot is defined. x stands for the height in a perspective plot.

(4) xd (coordinates of x-axis) and yd (coordinates of x-axis) indicate the position of the delta functions when the distribution of heat on tt(time) = 0 is represented by a set of the delta functions.

(5) The position for calculating the amount of heat is represented as ex and ey.

(6) The amount of heat on tt(time) = 1 at the position given by ex and ey is calculated and named zz. The command persp() shows a perspective plot.

(7) The amount of heat at tt(time) = 2 is calculated. Using this result, a new perspective plot is shown slightly above the perspective plot constructed in (6).

(8) The amount of heat at tt(time) = 3 is calculated. Using this result, a new perspective plot is shown slightly above the perspective plot constructed in (7).

(9) The amount of heat at tt(time) = 3 is calculated. Using this result, a new perspective plot is shown slightly above the perspective plot constructed in (8).

If zlim = c(0, 4) set in persp() in (6), (7), (8), and (9) is replaced with zlim = c(0, 1000), Fig. 2.22 (right) is shown.

2.15 IMPLICIT FUNCTIONS

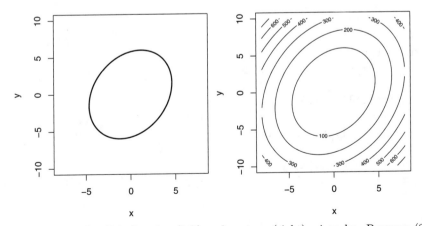

Figure 2.23 Implicit function (left) and contour (right), given by Program (2 - 38).

A function given by the form of an implicit function can be drawn. Fig. 2.23 (left) shows a curve that satisfies $5x^2 - 2xy + 3y^2 = 100$. Since Fig. 2.23 (right) shows a contour of z given by $z = 5x^2 - 2xy + 3y^2$, the curve in Fig. 2.23 (left) apparently satisfies $5x^2 - 2xy + 3y^2 = 100$. Fig. 2.23 is produced using Program (2 - 38).

Program (2 - 38)

```
function ()
{
# (1)
  elipse <- function(A, m, const, k){
    r <- A[1, 2]/sqrt(A[1, 1] * A[2, 2])
    Q <- matrix(0, 2, 2)
    Q[1, 1] <- sqrt(A[1, 1] %*% (1+r)/2)
    Q[1, 2] <- -sqrt(A[1, 1] %*% (1-r)/2)
    Q[2, 1] <- sqrt(A[2, 2] %*% (1+r)/2)
    Q[2, 2] <- sqrt(A[2, 2] %*% (1-r)/2)
    alpha <- seq(0, by = (2 * pi)/k, length = k)
    Z <- cbind(cos(alpha), sin(alpha))
    X <- t(m + const * Q %*% t(Z))
    X
  }
# (2)
  par(mfrow=c(1, 2), mai = c(1, 1, 0.2, 0.2),
    omi = c(0, 0, 0, 0))
# (3)
  bb <- matrix(c(5, -1, -1, 3), ncol = 2)
  aa <- solve(bb)
  mm <- c(0, 0)
  cc <-sqrt(100)
# (4)
  kk <- 5000
  xx <- elipse(aa, mm, cc, kk)
# (5)
  plot(xx[,1], xx[,2],type='n',xlab='x',ylab='y',
    xlim = c(-8, 8), ylim = c(-10, 10))
  lines(xx[,1], xx[,2], lwd = 2)
# (6)
  fun1 <- function(x, y){
    zz <- 5 * x^2 - 2 * x * y + 3 * y^2
  return(zz)
  }
# (7)
  nx <- 100
  ny <- 200
```

```
  xx <- seq(from = -8, to = 8, length = nx)
  yy <- seq(from = -10, to = 10, length = ny)
# (8)
  zz <- outer(xx, yy, fun1)
# (9)
  contour(xx, yy, zz, xlab = "x", ylab = "y")
}
```

(1) The function `elipse()` to derive the values which satisfy an implicit function with a quadrtic form is defined; "elipse" is a Spanish word for "ellipse". The original `elipse()` is found the following reference: Gráficos Estadísticos con R, Juan Carlos Correa y Nelfi González, Posgrado en Estadística, Universidad Nacional-Sede Medellín,
`http://cran.r-project.org/doc/contrib/grafi3.pdf`
(2) The graphics area is set.
(3) The constants for an implicit function with a quadratic form are given. These constants are used in the following implicit function with a quadratic form.

$$(\ x \quad y \ ) \begin{pmatrix} b_{11} & b_{12} \\ b_{21} & b_{22} \end{pmatrix} \begin{pmatrix} x \\ y \end{pmatrix} = c^2 \qquad (2.4)$$

`bb` is $\begin{pmatrix} b_{11} & b_{12} \\ b_{21} & b_{22} \end{pmatrix}$. `cc` stands for c. If `mm` is different from `c(0, 0)`, the resultant function moves in the direction of x or y, or both.
(4) After the number of calculation points on the curve that satisfy an implicit function is set as `kk`, `elipse()` derives the points that satisfy an implicit function. The resultant `xx` consists of two columns. The first one gives the values of x-axis. The second one gives the values of the y-axis.
(5) The elements of `xx` are drawn in a graph.
(6) The function `fun1()` for calculating $z = 5x^2 - 2xy + 3y^2$ is defined. This function is used for confirming that `elipse()` satisfies the implicit function.
(7) The values of the x-axis for calculating the value of `fun1()` are stored as `xx`. Those of the y-axis are stored as `yy`.
(8) The command `outer()` calculates the values of `func1()` at the grid points constructed using `xx` and `yy`; the results are stored as `zz`.
(9) The command `contour()` illustrates a contour using `zz`.

2.16   PROBABILITY DENSITY FUNCTIONS

A graph of a probability density function can also be drawn. Fig. 2.24 (left) given by Program (2 - 39) illustrates a normal distrution.

Program (2 - 39)
```
function() {
# (1)
```

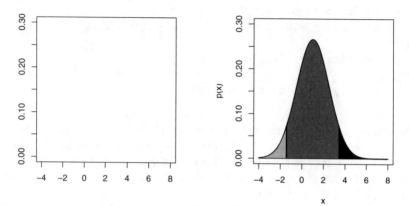

Figure 2.24   Probability density function of a normal distribution, given by Program (2 - 39).

```
  par(mai = c(1, 1, 1, 1), omi = c(0, 0, 0, 0))
# (2)
  plot(c(-4, 8), c(0, 0.3), xlab = "", ylab = "", type = "n")
# (3)
  fun1 <- function(x){
     yy <- dnorm(x, mean = 1, sd = 1.5)
  return(yy)
  }
# (4)
  curve(fun1, -4, 8, xlab = "x", ylab = "p(x)", lwd = 2,
    xlim = c(-4, 8), ylim = c(0, 0.3))
# (5)
  text(1, 0.1, "mean=1", cex = 1.2)
  text(1, 0.07, "sd=1.5", cex = 1.2)
}
```

(1) The graphics area is set.
(2) The coordinate axes are set and drawn.
(3) The function fun1() is defined. This function calculates the values of the probability density function of a normal distribution (the mean is 1 and the standard deviation is 1.5) using dnorm().
(4) The curve() draws the values given by fun1().
(5) The command text() adds letters in a graph.

Regions of a probability density function can be shown. Fig. 2.24 (right) given by Program (2 - 40) exemplifies it.

```
Program (2 - 40)
function() {
# (1)
```

```
  par(mai = c(1, 1, 1, 1), omi = c(0, 0, 0, 0))
# (2)
  xx <- seq(from = -4, to = 8, length = 200)
# (3)
  yy <- dnorm(xx, mean = 1, sd = 1.5)
  yyp <- pnorm(xx, mean = 1, sd = 1.5)
# (4)
  xxp1 <- xx[yyp <= 0.05]
  yyp1 <- yy[yyp <= 0.05]
  xxp1 <- c(xxp1, rev(xxp1))
  yyp1 <- c(yyp1, rep(0, length = length(yyp1)))
# (5)
  xxp2 <- xx[yyp > 0.05 & yyp < 0.95]
  yyp2 <- yy[yyp > 0.05 & yyp < 0.95]
  xxp2 <- c(xxp2, rev(xxp2))
  yyp2 <- c(yyp2, rep(0, length = length(yyp2)))
# (6)
  xxp3 <- xx[yyp >= 0.95]
  yyp3 <- yy[yyp >= 0.95]
  xxp3 <- c(xxp3, rev(xxp3))
  yyp3 <- c(yyp3, rep(0, length = length(yyp3)))
# (7)
  plot(c(-4, 8), c(0, 0.3), xlab = "x", ylab = "p(x)",
    type = "n")
# (8)
  lines(xx, yy, lwd = 2)
# (9)
  polygon(xxp1, yyp1, col = gray(0.8))
# (10)
  polygon(xxp2, yyp2, col = gray(0.5))
# (11)
  polygon(xxp3, yyp3, col = gray(0.2))
}
```

(1) The graphics area is set.

(2) The coordinates of the points where the values of the probability density function of a normal distribution are derived are stored as xx.

(3) The values of the probability density function of a normal distribution (the mean is 1 and the standard deviation is 1.5) at xx are calculated and named yy. The value of the cumulative distribution function (probability distribution function) of this normal distribution at xx is calculated and named yyp.

(4) Some elements of xx are extracted and named xxp1. These extracted values correspond to the elements of yyp that satisfy the condition that they are equal to or less than 0.05. Some elements of yy are extracted and named yyp1. These extracted values correspond to the elements of yyp that satisfy

the condition that they are equal to or less than 0.05. The command `rev()` inverts the elements of `xxp1`. The result is connected to the end of `xxp1` using `c()` and named `xxp1`. The values of 0s are connected to the end of `yyp1` to be called `yyp1`; the number of 0s is the same as that of elements of `yyp1`.

(5) `xxp2` and `yyp2` are produced in the same manner as those of (4).

(6) `xxp3` and `yyp3` are produced in the same manner as those of (4).

(7) The coordinate axes are drawn.

(8) The command `lines()` depicts a curve given by `xx` and `yy`; this curve shows the probability density function of a normal distribution.

(9) The area in which the value of the cumulative distribution function is equal to or less than 0.05 is painted using `xxp1` and `yyp1`. The argument `col = gray(0.8)` specifies light gray.

(10) The area in which the value of the cumulative distribution function is larger than 0.05 and less than 0.95 is painted using `xxp2` and `yyp2`. The argument `col = gray(0.5)` specifies intermediate gray.

(11) The area in which the value of the cumulative distribution function is larger than 0.95 is painted using `xxp3` and `yyp2`. The argument `col = gray(0.2)` specifies dark gray.

2.17   DIFFERENTIAL VALUES AND VALUES OF INTEGRALS

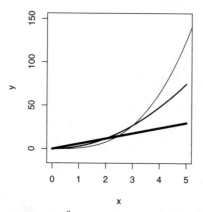

Figure 2.25   Graph of $y = x^3$ and its first derivatives and second derivatives, given by Program (2 - 41).

The results of differentiating a function analytically can be drawn in a graph. For example, Program (2 - 41) yields Fig. 2.25.

```
Program (2 - 41)
function() {
# (1)
```

```
  par(mai = c(1, 1, 1, 1), omi = c(0, 0, 0, 0))
# (2)
  curve(x^3, 0, 10, xlab = "x", ylab = "y", xlim = c(0, 5),
  ylim = c(-10, 150))
# (3)
  ff <- expression(x^3)
# (4)
  aa <- D(ff,"x")
# (5)
  fun1 <- substitute(curve(qq, 0, 5, add = T, xlab = "",
  ylab = "", lwd = 2), list(qq = aa))
# (6)
  eval(fun1)
# (7)
  bb <- D(D(ff,"x"), "x")
# (8)
  fun2 <- substitute(curve(qq, 0, 5, add = T, xlab = "",
  ylab = "", lwd = 4), list(qq = bb))
# (9)
  eval(fun2)
}
```

(1) The graphics area is set.
(2) The command curve() draws the graph of x^3.
(3) x^3 is transformed into an expression format and named ff.
(4) The command D() differentiates ff (i.e., x^3) with respect to x; the result function is named aa.
(5) The command substitute() uses aa (i.e., the derivatives of x^3 with respect to x) as an argument of curve(). The resultant function is named fun1.
(6) The command eval() conducts fun1 as a function.
(7) The command D() differentiates ff (i.e., x^3) twice with respect to x; the result is named bb.
(8) The command substitute() uses bb (i.e., the second derivatives of x^3 with respect to x) as an argument of curve(). The resultant function is named fun2.
(9) The command eval() conducts fun2 as a function.

A function of many variables is also differentiated partially in analytical form. Its use allows users to illustrate the behavior of the method of steepest descent, as in Fig. 2.26 (left). The trajectories of the two variables are apparently orthogonal to the contour lines. Program (2 - 42) illustrates this graph.

Program (2 - 42)
```
function() {
# (1)
```

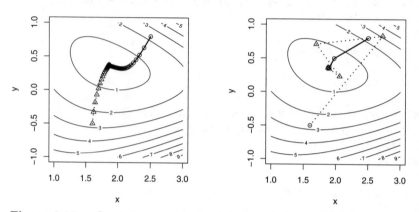

Figure 2.26   Graph for showing the behavior of the method of steepest descent, given by Program (2 - 42) (left). Graph for showing the behavior of the Newton-Raphson method, given by Program (2 - 43) (right).

```
  par(mai = c(1, 1, 1, 1), omi = c(0, 0, 0, 0))
# (2)
  func1 <- function(x,y){
    zz <- (x - 2)^2 + (y - 1)^2 + 0.5 * x^2 * y^2
  return(zz)
  }
# (3)
  nx <- 101
  xx <- seq(from = 1, to = 3, length = nx)
  ny <- 201
  yy <- seq(from = -1, to= 1, length = ny)
# (4)
  xymat <- outer(xx, yy, func1)
  contour(xx, yy, xymat, xlab = "x", ylab = "y")
# (5)
  ff <- expression((x - 2)^2 + (y - 1)^2 + 0.5 * x^2 * y^2)
# (6)
  fx <- D(ff, "x")
# (7)
  funx <- function(x,y){}
  body(funx) <- fx
# (8)
  fy <- D(ff, "y")
# (9)
```

```
funy <- function(x,y){}
body(funy) <- fy
# (10)
  xx <- 2.5
  yy <- 0.8
  xx2 <- xx
  yy2 <- yy
  points(xx, yy, pch=1)
# (11)
  for(ii in 1:100){
    xx <- xx - funx(xx, yy) * 0.04
    yy <- yy - funy(xx, yy) * 0.04
    points(xx, yy, pch = 1)
    lines(c(xx2, xx), c(yy2, yy), lwd=2)
    xx2 <- xx
    yy2 <- yy
  }
# (12)
  xx <- 1.6
  yy <- -0.5
  xx2 <- xx
  yy2 <- yy
  points(xx, yy, pch=2)
  for(ii in 1:100){
    xx <- xx - funx(xx, yy) * 0.04
    yy <- yy - funy(xx, yy) * 0.04
    points(xx, yy, pch = 2)
    lines(c(xx2, xx), c(yy2, yy), lwd = 2, lty = 3)
    xx2 <- xx
    yy2 <- yy
  }
}
```

(1) The graphics area is set.
(2) The function func1() with two variables (x and y) is defined.
(3) The values of the positions of the x-axis for drawing a contour are stored as xx. Those of the y-axis are stored as yy. The elements of xx are $(1.00, 1.02, 1.04, \ldots, 3.00)$. The elements of yy are $(-1.00 - 0.99 - 0.98, \ldots, 1.00)$.
(4) The command outer() calculates the values of func1() at the grid points constructed using xx and yy; the results are stored as zz. Then, contour() illustrates the contour of xymat.
(5) $(x - 2)\char94 2 + (y - 1)\char94 2 + 0.5 * x\char94 2 * y\char94 2$ is converted to an expression format and named ff.

(6) The command D() differentiates ff (i.e., (x - 2)^2 + (y - 1)^2 + 0.5 * x^2 * y^2) with respect to x. The result is transformed into an expression format and named fx.

(7) The function funx() is made to do nothing. The command body() copies the content of fx to funx().

(8) The command D() differentiates ff (i.e., (x - 2)^2 + (y - 1)^2 + 0.5 * x^2 * y^2) with respect to y. The result is transformed into an expression format and named fy.

(9) The function funy() is made to do nothing. The command body() copies the content of fy to funy().

(10) The initial value of xx is set at 2.5. That of yy is set at 0.8. These two values are also stored as xx2 and yy2, respectively. The command points() plots the point given by the initial values in the contour.

(11) xx and yy are changed when moving the point given by xx and yy along the gradient at the point; the direction is set to move the points to the place where the value of func1() is less than that at the current place. 0.04 gives the length of the changes of xx and yy. The command for(ii in 1:100) repeats this task 100 times.

(12) The initial value of xx is set at 1.6. That of yy is set at −0.5. Steps (10) and (11) are carried out.

A function of many variables is also partially differentiated twice or more in analytical form. This method allows users to illustrate the behavior of the Newton-Raphson method, as in Fig. 2.26 (right). This graph shows that the searching point of the Newton-Raphson method moves by a large margin in some occasions. Program (2 - 43) draws this graph.

Program (2 - 43)

```
function() {
# (1)
  par(mai = c(1, 1, 1, 1), omi = c(0, 0, 0, 0))
# (2)
  func1 <- function(x,y){
    zz <- (x - 2)^2 + (y - 1)^2 + 0.5 * x^2 * y^2
  return(zz)
  }
# (3)
  nx <- 100
  xx <- seq(from = 1, to = 3, length = nx)
  ny <- 200
  yy <- seq(from = -1, to = 1, length = ny)
# (4)
  xymat <- outer(xx, yy, func1)
  contour(xx, yy, xymat, xlab = "x", ylab = "y")
# (5)
  ff <- expression((x - 2)^2 + (y - 1)^2 + 0.5 * x^2 * y^2)
```

```
# (6)
  fx <- D(ff, "x")
  funx <- function(x,y){}
  body(funx) <- fx
# (7)
  fy <- D(ff, "y")
  funy <- function(x,y){}
  body(funy) <- fy
# (8)
  fxx <- D(D(ff, "x"), "x")
  funxx <- function(x,y){}
  body(funxx) <- fxx
# (9)
  fxy <- D(D(ff, "x"), "y")
  funxy <- function(x,y){}
  body(funxy) <- fxy
# (10)
  fyy <- D(D(ff, "y"), "y")
  funyy <- function(x,y){}
  body(funyy) <- fyy
# (11)
  xx <- 2.5
  yy <- 0.8
  xxyy <- c(xx, yy)
  points(xxyy[1], xxyy[2], pch = 1)
  xxyy2 <- xxyy
# (12)
  for(ii in 1:100){
    hes1 <- matrix(c(funxx(xxyy[1], xxyy[2]), funxy(xxyy[1],
     xxyy[2]), funxy(xxyy[1], xxyy[2]), funyy(xxyy[1],
     xxyy[2])), ncol = 2)
    ihes1 <- solve(hes1)
    dxxyy <- c(funx(xxyy[1], xxyy[2]), funy(xxyy[1], xxyy[2]))
    xxyy <- xxyy - ihes1 %*% dxxyy
    points(xxyy[1], xxyy[2], pch=1)
    lines(c(xxyy2[1], xxyy[1]), c(xxyy2[2], xxyy[2]), lwd = 2)
    xxyy2 <- xxyy
  }
# (13)
  xx <- 1.6
  yy <- -0.5
  xxyy <- c(xx, yy)
  points(xxyy[1], xxyy[2], pch = 1)
  xxyy2 <- xxyy
  for(ii in 1:100){
```

```
    hes1 <- matrix(c(funxx(xxyy[1], xxyy[2]), funxy(xxyy[1],
      xxyy[2]), funxy(xxyy[1], xxyy[2]), funyy(xxyy[1],
      xxyy[2])), nrow = 2)
    ihes1 <- solve(hes1)
    dxxyy <- c(funx(xxyy[1], xxyy[2]), funy(xxyy[1], xxyy[2]))
    xxyy <- xxyy - ihes1 %*% dxxyy
    points(xxyy[1], xxyy[2], pch = 2)
    lines(c(xxyy2[1], xxyy[1]), c(xxyy2[2], xxyy[2]),
      lwd = 2, lty = 3)
    xxyy2 <- xxyy
  }
}
```

(1) The graphics area is set.

(2) The function func1() with two variables (x and y) is defined.

(3) The points of the x-axis for deriving the values of func1() and for illustrating a contour are given as xx. Those of the y-axis are given as yy. Since seq(from = 1, to = 3, length = nx) indicates nx equally spaced numbers between 1 and 3, the elements of xx are $(1.00, 1.02, 1.04, \ldots, 3.00)$. The elements of yy are $(-1.00 - 0.99 - 0.98, \ldots, 1.00)$ by the same logic.

(4) The command outer() calculates the values of func1() at the grid points constructed using xx and yy; the results are stored as xymat. Then, contour() draws the contour of xymat.

(5) (x - 2)^2 + (y - 1)^2 + 0.5 * x^2 * y^2 is converted to an expression format and named ff.

(6) The command D() differentiates ff (i.e., (x - 2)^2 + (y - 1)^2 + 0.5 * x^2 * y^2) with respect to x; the result is called funx.

(7) The command D() differentiates ff (i.e., (x - 2)^2 + (y - 1)^2 + 0.5 * x^2 * y^2) with respect to y; the result is called funy.

(8) The command D() differentiates ff (i.e., (x - 2)^2 + (y - 1)^2 + 0.5 * x^2 * y^2) twice with respect to x; the result is called funxx.

(9) The command D() differentiates ff (i.e., (x - 2)^2 + (y - 1)^2 + 0.5 * x^2 * y^2) with respect to x and differentiates the resultant function with respect to y; the result is called funxy.

(10) The command D() differentiates ff (i.e., (x - 2)^2 + (y - 1)^2 + 0.5 * x^2 * y^2) twice with respect to y; the result is called funyy.

(11) The initial value of xx is set at 2.5. The initial value of yy is set at 0.8. These values are also stored as xx2 and yy2, respectively. The command points() plots the point given by the initial values in the contour.

(12) The update by the Newton-Raphson method is conducted once at the point given by xx and yy. If the argument of solve() is a square matrix only, the command solve() outputs an inverse matrix. The command for(ii in 1:100) repeats the task 100 times.

(13) The initial value of xx is set at 1.6. The initial value of yy is set at -0.5. Then, the same tasks as (11) and (12) are carried out.

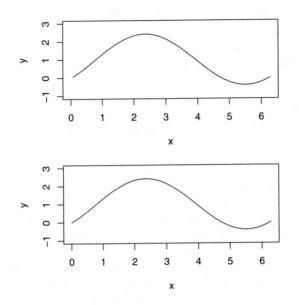

Figure 2.27   Result of numerical integration (top) and that of analytical integration (bottom) given by Program (2 - 44).

An R command enables numerical integration. Fig. 2.27 Program (2 - 44) compares the result of numerical integration with that of analytical integration.

Program (2 - 44)

```
function() {
# (1)
  par(mfrow=c(2, 1), mai = c(1, 1, 1, 1), omi = c(0, 0, 0, 0))
# (2)
  fun1 <- function(tt){
    uu <- sin(tt) + cos(tt)
  return(uu)
  }
# (3)
  fun2 <- function(tt){
    uu <- -cos(tt) + sin(tt) + 1
    return(uu)
  }
# (4)
  xx <- seq(from= 0.02 * pi, to = 2 * pi, length = 100)
  yy <- rep(0,100)
  for(ii in 1:100){
```

```
    yy[ii] <- integrate(fun1, 0, xx[ii])$value
  }
# (5)
  plot(xx, yy, xlab = "x", ylab = "y", type = "n", xlim=c(0, 2 * pi),
  ylim=c(-1, 3))
# (6)
  lines(xx, yy)
  curve(fun2, 0, 2*pi, xlab = "x", ylab = "y", xlim=c(0, 2*pi),
  ylim=c(-1, 3))
}
```

(1) The graphics area is set.
(2) The function fun1() (i.e., sin(tt) + cos(tt)) is defined.
(3) The function fun2() (i.e., -cos(tt) + sin(tt) + 1) is defined. The indefinite integral of fun1() is carried out. The integral constant is derived to determine the value of the function 0 when the value of the variable is 0.
(4) The command integrate() conducts numerical integration of fun1(). The minimum of the integral range is always 0. Its maximum is set at $0.02, 0.04, 0.06, \ldots, 2\pi$. These results are stored as yy.
(5) The coordinate axes are drawn.
(6) The command lines() draws the values of yy. The command curve() draws the values given by fun2().

2.18 SUMMARY

1. The command stem() displays a stem-and-leaf display.

2. The command hist() draws a histogram.

3. The command hist() needs preprocessed data given by floor().

4. The density() illustrates a smooth probability density function.

5. The command stripchart() draws a strip chart.

6. The command boxplot() illustrates a boxplot.

7. The command par(new = T) describes that the current graph will remain as it is and a new graph will be superimposed.

8. The command par(new = T) allows users to draw both a bar plot and a line chart in a graph.

9. The command par(new = T) realizes a graph for comparing two pie charts.

10. The command axis() draws a coordinate axis.

11. The command `mtext()` adds explanations of a coordinate axis.

12. If the command `par()` resets the graphics area, a new coordinate axis can be illustrated at any place.

13. The command `barplot()` constructs a population pyramid.

14. The command `arrows()` draws confidence intervals.

15. The command `pairs()` draws a matrix of scatter plots.

16. The command `stars()` illustrates a radar chart.

17. The division of a graphics window by `layout()` realizes a histogram along a coordinate axis.

18. The command `curve()` draws a graph of a function with one variable.

19. The command `curve()` draws a probability density function.

20. An R function is defined in a recursive manner.

21. The command `contour()` draws a contour.

22. The command `filled.contour()` illustrates a contour with shading.

23. The command `persp()` illustrates a perspective plot.

24. The command `persp()` realizes a bird's-eye-view map.

25. The command `image()` displays the values with two variables using a color code. This command also draws maps.

26. The command `D()` differentiates a function analytically.

27. The command `integrate()` conducts numerical integration of a function.

EXERCISES

2.1 To ensure that the correct histogram is obtained, replace the part of (2) in Program (2 - 7)(70 page) with:

```
# (2)'
  xx <- c(6, 10, 13.99999999, 14, 15.999999)
```

2.2 Replace (2) in Program (2 - 8)(page 71) with the following program. 10000 elements of xx are extracted from the population that obeys a normal distribution in which the mean is −1 and the standard deviation is 5. Ensure that the resultant probability density function is similar to the probability density function that the population obeys.

```
# (2)'
  set.seed(103)
  xx <- rnorm(10000, mean = -1, sd = 5)
```

2.3   To draw a graph of a contingency table with three variables using mosaicplot(), replace (2)(3) in Program (2 - 9)(page 72) with:

```
# (2)'
  xx <- matrix(c(2, 2, 3, 4), ncol = 2)
  yy <- matrix(c(2, 4, 3, 6), ncol = 2)
  zz <- matrix(c(8, 2, 3, 6), ncol = 2)
  darray <- array(rep(0, 2 * 2 * 3), c(2, 2, 3),
    dimnames = list(c("a1", "a2"),  c("b1", "b2"),
    c("c1", "c2", "c3") ))
  darray[,,1] <- xx
  darray[,,2] <- yy
  darray[,,3] <- zz
# (3)'
  mosaicplot(darray, color = rainbow(3), main = "")
```

2.4   To thin the lines, add boxwex = 0.5 in boxplot() located in (3) in Program (2 - 11)(page 76).

2.5   To alter the shape of the boxes, add notch = T in boxplot() located in (3) in Program (2 - 11)(page 76).

2.6   To draw an x-axis on the top side of the graph, replace axis(3) in (6) of Program (2 - 13)(page 80) with axis(4).

2.7   To add explanations of y-axes, add mtext("y1", side = 2, line = 2) at the end of (5) in Program (2 - 16)(page 84) and add mtext("y2", side = 2, line = 2) at the end of (12).

2.8   To confirm that the area of data-a is not segmentized, replace (8) and (9) in Program (2 - 17)(page 86).

```
# (8)'
  name2 <- c("", "data-b1", "data-b2", "data-c1", "data-c2")
# (9)'
  pie(yy2, labels = name2, col =c("red", "gold", "blue",
    "gold", "blue"), lty = 0)
```

2.9   Modify the age groups in Fig. 2.9 (Program (2 - 18)(page 87)), which is a population pyramid by single year of age, to classify males and females in such a way that the first bar indicates the numbers of males and females between 0 and 4, the second bar indicates the numbers of males and females between 5 and 9, the third bar indicates the numbers of males and females between 10 and 14, etc.

2.10   Replace histograms with boxplots in Fig. 2.10 (Program (2 - 19)(page 89)) to depict the distribution of xx as one variable and that of yy as another variable.

2.11   To add sharp edges of the arrowheads, add the following program at the end of Program (2 - 20)(page 91).

```
# (6)
  sd2 <- c(80, 100, 120, 60)
  arrows(xp, yy - sd2, xp, yy + sd2, angle = 40, length = 0.1)
  arrows(xp, yy + sd2, xp, yy - sd2, angle = 40, length = 0.1)
```

2.12   To add histograms, (5) in Program (2 - 23)(page 94) is replaced with the following program. The function panel.hist() is included in the manual of pairs().

```
# (5)'
  panel.hist <- function(x, ...)
  {
    usr <- par("usr")
    on.exit(par(usr))
    par(usr = c(usr[1:2], 0, 1.5) )
    h <- hist(x, plot = FALSE)
    breaks <- h$breaks
    nB <- length(breaks)
    y <- h$counts
    y <- y/max(y)
    rect(breaks[-nB], 0, breaks[-1], y, col="cyan", ...)
  }
  pairs(data1, pch = pch1, cex.labels = 3, cex.axis = 2, cex = 3,
    diag.panel = panel.hist)
```

2.13   To construct a graph for comparing xx1, xx2, and xx3, replace (3) of Program (2 - 24)(page 95) with:

```
# (3)'
  stars(data1, locations = c(0,0), scale = F, draw.segments = TRUE,
    col.segments = 0, key.loc = c(0, 0))
```

2.14   Delete (3) in Program (2 - 26)(page 97) and replace (4) with (4)';
it yields the same graph.

```
# (4)'
  curve(0.05 * (x - 7)^2 - 1, 2, 9, add = T, lwd = 4)
```

2.15   Modify Program (2 - 26)(page 97) to draw the graph of log() and tan().

2.16 Modify Program (2 - 28)(page 100) for writing letters in the graph using `text()`.

2.17 Replace `filled.contour()` in Program (2 - 29)(page 100) with `image()`.

2.18 Alter the arguments of `persp()` in (6) of Program (2 - 30)(page 101). For example, `r =`, `expand =`, `ltheta =`, and `lphi =` are added and diverse values are set .

2.19 Replace `persp()` in Program (2 - 31)(page 102) with `image()`. For this, delete (8) and add `add = T` in `contour()`. That is, the following program is used.

```
# (7)'
  image(zz, breaks = seq(from = -1, to = 1, by = 0.2),
    col = rainbow(10))
# (8)'
# (9)'
  contour(zz, add = T, xlab = "x", ylab = "y", levels =
    seq(from = -1, to = 1, by = 0.2))
```

2.20 Replace (3) in Program (2 - 31)(page 102) with the following program. The contour turns into pastel rainbows. Change the value of `pas1` between 0 and 1 to observe the change in the color tone.

```
# (3)'
  colfunc1 <- function(x) {
    xave <- (x[-1, -1] + x[-1, -(ncol(x) - 1)] + x[-(nrow(x) -1),
      -1] + x[-(nrow(x) -1), -(ncol(x) - 1)]) / 4
    col1 <- rainbow(10)
    pas1 <- 0.6
    col2 <- round(pas1 * 255 + (1 - pas1s) * col2rgb(col1))
    col3 <- rgb(col2[1,], col2[2,], col2[3,], max=255)
    colors <-  col3[cut(xave, breaks = seq(from = -1, to = 1,
      by = 0.2), include.lowest = T)]
    return(colors)
  }
```

2.21 Modify (1) of Program (2 - 34)(page 108) to alter the use of color in the perspective plot.

2.22 Replace (7) of Program (2 - 34)(page 108) with the following R program. An advertising display shows up on the curved surface on the perspective plot. The command `trans3d()` converts coordinates in a 3-dimensional space into coordinates in the perspective plot.

```
# (7)'
  p1 <- persp(xx, yy, zz, theta = 25, phi = 60, ltheta = 70,
```

```
 lphi = 20, shade = 0.0, col = colfunc3(map2),r = 100,
  border = F)
xc <- c(2, 2.5)
yc <- c(6.5, 7)
xy1 <-  trans3d(xc, yc,fun1(xc, yc), pmat = p1)
xy2 <-  trans3d(xc, yc,fun1(xc, yc) + 1, pmat = p1)
xrec <- c(xy1$x, rev(xy2$x))
yrec <- c(xy1$y, rev(xy2$y))
polygon(xrec, yrec, col = "skyblue")
xd <- 2.25
yd <- 6.75
xy3 <-  trans3d(xd, yd, fun1(xd, yd) + 0.5, pmat = p1)
text(xy3, "map", srt = -23)
```

2.23 Change the use of color in Fig. 2.19 (Program (2 - 35)(page 110)). Furthermore, change the positions of letters.

2.24 Rotate the perspective plot in Fig. 2.20(Program (2 - 35)(page 110)). Change the use of lights.

2.25 If theta = 30 in persp() in (7) of Program (2 - 36)(page 113) is substututed with theta = 210, the back side of the object is displayed. However, the colors of the back side of the bars are not correct. Improve Program (2 - 44)to display correct colors.

2.26 Modify Program (2 - 39)(page 121) to draw a Poisson distribution (the mean of the Poisson distribution is 3). Use dpois().

2.27 Replace (6) in Program (2 - 41)(page 124) with the following (6)'. In this method, fun11() is made to do nothing at first. Then, body() copies the content of fun1 to fun11(). Do the same thing with (9).

```
# (6)'
  fun11 <- function(){}
  body(fun11) <- fun1
  fun11()
```

2.28 Modify Fig. 2.26 (left) (Program (2 - 42)(page 125)) to illustrate the behaviors given by other initial values.

2.29 Modify Program (2 - 44)(page 131) to draw a graph showing that the result of numerical integration of $\exp(2x)$ is close to the function $0.5\exp(2x)$.

CHAPTER 3

INTERACTIVE R PROGRAMS

3.1 INTRODUCTION

There is high demand for the use of R in statistical analysis and graphical
representation without typing R commands one after another or producing
R programs. Hence, to allow a lot more people to use R, we need to develop favorable environments for using R without considering the presence of
R commands or R programs. Interactive R programs provide one solution.
When we use interactive programs based on Visual Basic or Visual C++, we
can carry out statistical analysis and graphic representation without paying
any attention to the running software. In the same way, if we have interactive
R programs, we can perform statistical analysis and graphical representation
using whole functions of R without being aware of R commands or R programs.

However, R was developed for the purpose of statistical analysis and graphical representation by inputting R commands one by one using a keyboard.
The use of R as R programs is a secondary matter. The production of interactive programs has received little attention. Therefore, good looks and
user-friendliness given by Visual Basic or Visual C++ cannot be expected for

interactive R programs. However, if we take into account the great significance of allowing R to be used without assuming any knowledge of R commands or R programs, good looks and user-friendliness matter little. Furthermore, in the procedures of statistical analysis and graphical representation, we seldom cope with diverse data files, complex condition settings, and tangled conditional branchings based on calculations. Hence, a complex graphical interface is not a requisite. If statistical analysis and graphical representation need complicated procedures, users can store the results of an interactive program as a data file to be used with another interactive program. The latter interactive program can be chosen according to the results given by the former interactive program. This strategy completes an assigned task using R programs without advanced interfaces.

Several simple interactive R programs are introduced below. These descriptions provide readers with a synopsis of basic techniques for using R in this manner. Although R programs serve as examples aiming to fulfill a specific function, the techniques in such programs can be used widely when users develop interactive R programs. The author expects that many readers will develop interactive R programs using all these methods.

3.2 POSITIONING BY MOUSE ON A GRAPHICS WINDOW

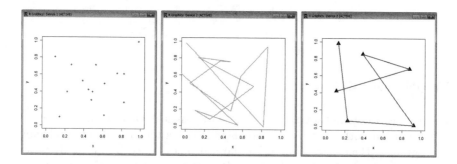

Figure 3.1   Points plotted by Program (3 - 1)(left). Broken line drawn by Program (3 - 2)(middle). Points and broken line illustrated by Program (3 - 3) (right).

In interactive R programs, it is useful to extract information on the location of the mouse by clicking it in graphics windows; the information will be used later on. This technique is easier than the use of the keyboard.

Fig. 3.1 (left) exemplifies a graph illustrated by clicking the mouse. Program (3 - 1) enables this work.

Program (3 - 1)
```
function() {
# (1)
```

```
  plot(c(0, 1), c(0, 1), xlim = c(0, 1), ylim = c(0, 1),
    type = "n", xlab = "x", ylab = "y")
# (2)
  repeat{
# (3)
    locator(type = "p", pch = 16, col = "magenta")
  }
}
```

(1) Coordinate axes are drawn.
(2) The command **repeat** indicates that the procedure in parentheses repeats. Users press the escape key to stop the repetition.
(3) If the mouse is clicked in the graphics window, **locator()** detects the position of the mouse. The argument **type = "p"** specifies that a symbol is drawn at a place; the type of symbol is indicated by **pch =**. The argument **col =** sets the color of the symbol.

Fig. 3.1(middle), given by Program (3 - 2), is an example of a graph obtained by clicking the mouse.

Program (3 - 2)
```
function() {
# (1)
  plot(c(0, 1), c(0, 1), xlim = c(0, 1), ylim = c(0, 1),
    type = "n", xlab = "x", ylab = "y")
# (2)
  repeat{
# (3)
    locator(type = "l", col = "green", lwd = 3)
  }
}
```

The argument **type = "l"** specifies that the points where the mouse is clicked are connected by straight lines.

Fig. 3.1 (right) shows the points and the broken line drawn by detecting the points where the mouse is clicked to be stored as numerical data. This figure is jsillustrated using Program (3 - 3).

Program (3 - 3)
```
function() {
# (1)
  plot(c(0, 1), c(0, 1), xlim = c(0, 1), ylim = c(0, 1),
    type = "n", xlab = "x", ylab = "y")
# (2)
  pxx2 <- -1
# (3)
  repeat{
# (4)
```

```
      repeat{
# (5)
         p1<- locator(1)
         pxx <- p1$x
         pyy <- p1$y
         if(pxx >= 0 && pxx <= 1 && pyy >= 0 && pyy <= 1) break
      }
# (6)
         points(pxx, pyy, pch = 17, col = "blue", cex = 2)
# (7)
         if(pxx2 > 0){
           lines(c(pxx, pxx2), c(pyy, pyy2), col = 2, lwd = 3)
         }
# (8)
         pxx2 <- pxx
         pyy2 <- pyy
      }
}
```

(1) Coordinate axes are drawn.

(2) pxx2 is set to be −1. Since no previous points are at hand when the first point is plotted, this procedure is required in order to prevent drawing a straight line through the first point.

(3) In a graphics window, points where the mouse is clicked are plotted to be connected by straight lines. The command **repeat** repeats this procedure.

(4) The command **repeat** repeats the procedure in parentheses. The repetition is stopped by executing **break**.

(5) In a graphics window, the points where the mouse is clicked are detected by locator() and the result is stored in p1. The value of the x-axis at the point is stored as p1$x and that of the y-axis is stored as p1$y; they are put in pxx and pyy, respectively. If pxx and pyy do not satisfy the condition that both values are equal to or larger than 0, and equal to or less than 1, **repeat** the work until the next click

(6) The command **points()** plots a point at a point where the value of the x-axis is pxx and that of the y-axis is pyy.

(7) The command **lines()** connects the point plotted in (6) and the previous point with a straight line. If the point plotted in (6) is the first one, no straight lines are drawn. For this purpose, pxx2 is set to be −1. Since this setting does not satisfy the condition of pxx2 > 0, a straight line is not drawn.

(8) The location of the points plotted in (6) is stored as pxx2 and pyy2.

Program (3 - 4)

```
function() {
# (1)
  plot(c(0, 1), c(0, 1), xlim = c(0, 1), ylim = c(0, 1),
    type = "n", xlab = "x", ylab = "y")
```

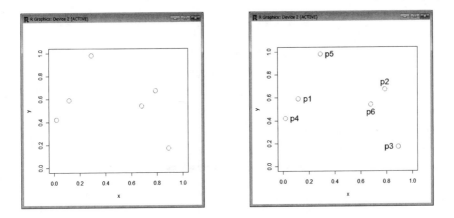

Figure 3.2   Points for adding explanations (right). Added explanations using Program (3 - 4) (left).

```
# (2)
  xx <- c(0.12, 0.79, 0.89, 0.02, 0.29, 0.68)
  yy <- c(0.59, 0.67, 0.17, 0.42, 0.98, 0.54)
# (3)
  points(xx, yy, col = "red", cex = 2)
# (4)
  ch1 <- c("p1", "p2", "p3", "p4", "p5", "p6")
  identify(x = xx, y = yy, labels = ch1, cex = 1.5, col = "blue")
}
```

(1) Coordinate axes are drawn.

The values of the coordinates for plotting points are given. The values of the x-axis are stored in **xx**. Those of the y-axis are stored in **yy**.

(3) The points specified in (2) are plotted.

(4) The explanations of the six points are set in **ch1**. When the mouse is clicked at a point close to one of the points that **xx** and **yy** indicate, the command **identify()** draws a text among **labels = ch1** sequentially. If the mouse is clicked more than or equal to 0.25 inch (about 6.35 mm) away from the points specified by **xx** and **yy** or the mouse is clicked near the point already explained, no explanations are added.

3.3   INPUTTING VALUES ON THE CONSOLE WINDOW TO DRAW A GRAPH

If appropriate R programs are provided, users of R do not have to pay attention to the presence of R programs when they carry out the procedure: the input of a numerical value in the console window, calculation using the value,

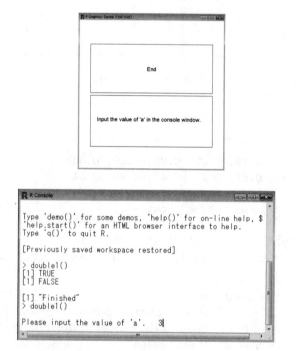

Figure 3.3   Selection of task in the process of Programs (3 - 5) and (3 - 6) (top). Input of the value of "a" in the process of Programs (3 - 5) and (3 - 6) (bottom).

and the display of a graph of the calculation result. Fig. 3.3 and Fig. 3.4 illustrate the processes of operating such an R program. A task is selected by clicking the mouse in a graphics window (Fig. 3.3 (left)). A numerical value is input in a console window (Fig. 3.3 (right)). The result of the calculation is diplayed as a graph (Fig. 3.4). This sequence of actions is realized using two R programs: Program (3 - 5) and Program (3 - 6).

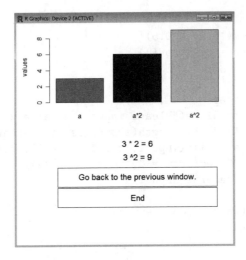

Figure 3.4 Display of the result of the interactive task by Programs (3 - 5) and (3 - 6).

Program (3 - 5)
```
function() {
# The name of this R program should be double1().
# (1)
  exit1 <- function() {
    opt <- options(show.error.messages = F)
    on.exit(options(opt))
    stop()
  }
# (2)
  graphics.off()
# (3)
  options(locatorBell = FALSE)
# (4)
  par(mfrow = c(1, 1), mai = c(0.1, 0.1, 0.1, 0.1),
    omi = c(0.1, 0.1, 0.1, 0.1))
  plot(c(0, 1), c(0, 1), xlab = "", ylab = "", type = "n",
    axes = F)
```

```
    text(0.5, 0.7, labels = "End", col = 1, cex = 1.3)
    rect(0.01, 0.51, 0.99, 0.89)
    text(0.5, 0.3, labels = "Input the value of 'a' in the
      console window.    ", col = 1, cex = 1.3)
    rect(0.01, 0.09, 0.99, 0.49)
# (5)
    p1<- locator(1)
# (6)
    if(p1$y > 0.1 && p1$y < 0.5){
      dev.off(which = dev.prev())
# (7)
      repeat{
        writeLines(sep = "\n","")
        input1 <- readline("Please input the value of 'a'.    ")
        if(input1 != "" && length(input1[is.na(charmatch
          ( unlist(strsplit(input1, split = NULL)), c(as.character
          ( c(".","-",seq(from = 0, to = 9))  ))))]) == 0){
          aa <- as.numeric(input1)
# (8)
          dev.set(which = dev.next())
          double2(aa)
        }
        else {
          print("Please input again.")
        }
      }
    }
# (9)
    else{
print(dev.interactive())
      dev.off(which = dev.prev())
print(dev.interactive())
      writeLines(sep = "\n","")
      print("Finished")
      exit1()
    }
}
```

(1) The function of exit1() is defined. This function deters the display of error messages when forced quit is executed. The command options() specifies the fundamental setting for calculation and display using R, The command show.error.messages = FALSE indicates that error messages are not output. This setting is named opt. The command on.exit(options(opt)) deters the output of an error message when forced quit is executed.
(2) The command graphics.off() closes all graphics windows.

(3) The command `options(locatorBell = FALSE)` specifies the setting of turning off the beeping sound.

(4) The coordinate axes in a graphics window are set. A selection on a graphics window is prompted (Fig. 3.3 (left)).

(5) The command `locator(1)` detects the position of the point where the mouse is clicked in a graphics window. The values are stored in `p1`. `p1$x` is the value of the x-axis. `p1$y` is the value of the y-axis.

(6) If the value of `p1$y` is larger than 0.1 and less than 0.5, `dev.off(which = dev.prev())` closes the graphics window.

(7) The command `readline()` allows the input of the value of "a" in a console window (Fig. 3.3 (right)). If the input(`input1`) contains something other than numerical values, subtraction sign, and decimal points, `repeat` works to prompt another input. If `input1` is an appropriate one, `as.numeric()` transforms `input1` into a numerical value to be stored as `aa`.

(8) The next window is opened. The R program `double2()` with the argument `aa` is performed (Fig. 3.4).

(9) The command `dev.off(which = dev.prev())` closes the graphics window. The command `writeLines(sep = "\n","")` inserts a line feed in a console window. The command `print("Finished")` displays `print"Finished"` in a console window. The command `exit1()` terminates the task. An error message is not displayed.

Program (3 - 5) uses the R program `double2()`: Program (3 - 6).

Program (3 - 6)

```
function (aa)
{
# The name of this R program should be double2().
# (1)
  exit1 <- function() {
    opt <- options(show.error.messages = F)
    on.exit(options(opt))
    stop()
  }
# (2)
  bringToTop()
# (3)
  par(mfrow = c(2, 1), mai = c(0.8, 0.8, 0.1, 0.1),
    omi = c(0.1, 0.1, 0.1, 0.1))
# (4)
  barplot(c(aa, aa*2, aa*aa), names = c("a","a*2","a^2"),
    col = c("red", "blue", "green"), horiz = F, border = "black",
    ylab = "values")
# (5)
  plot(c(0, 1), c(0, 1), xlab = "", ylab = "", type = "n",
    axes = F)
```

```
text1 <- paste(as.character(aa), " * 2 = ", as.character(aa*2),
  sep = "")
text(0.5, 0.95, labels = text1, col = 1, cex = 1.3)
text1 <- paste(as.character(aa), " ^2 = ", as.character(aa^2),
  sep = "")
text(0.5, 0.75, labels = text1, col = 1, cex = 1.3)
# (6)
text(0.5, 0.45, labels = "Go back to the previous window.",
  col = 1, cex = 1.3)
rect(0.01, 0.31, 0.99, 0.59)
text(0.5, 0.15, labels = "End", col = 1, cex = 1.3)
rect(0.01, 0.01, 0.99, 0.29)
# (7)
p1<- locator(1)
# (8)
if(p1$y > 0.0 && p1$y < 0.3){
  dev.off(which = dev.prev())
  writeLines(sep = "\n", "")
  print("Finished")
  exit1()
}
# (9)
else{
  dev.off(which = dev.prev())
  dev.set(which = dev.next())
  double1()
}
}
```

The function exit1() for deterring the display of error messages is defined.
(2) The command bringToTop() displaces the working window to the front.
(3) The coordinate axes in a graphics window are set.
(4) The value of aa, twice the value of aa, and the squared value of aa are represented as a bar chart.
(5) The value of aa, twice the value of aa, and the squared value of aa are represented as numbers using text().
A selection on a graphics window is prompted.
(7) The command locator(1) detects the position of the point where the mouse is clicked in a graphics window. The value is stored in p1.
(8) If p1$y is larger than 0.1 and less than 0.3, dev.off(which = dev.prev()) closes the graphics window. The command writeLines(sep = "\n","") inserts a line feed in a console window. The command print("Finished") displays print"Finished" in a console window. The command exit1() terminates the task. An error message is not displayed.

(9) If `p1$y` does not satisfy the condition that it is larger than 0.1 and less than 0.3, `dev.off(which = dev.prev())` closes the graphics window. The command `dev.set(which = dev.next())` opens a new graphics window. The function `double1()` is executed.

To produce interactive R programs, manageable R programs are constructed easily if one program is composed of a set of R programs to perform a series of tasks. If each R program takes a limited task to be executed when necessary, the selection of one task to perform is realized by describing the commands to execute one R program. This makes the structure of the program interpretable. Furthermore, when a previous task is executed again, for example, another input of a number due to an inappropriate earlier value, the task is completed by executing the R program to input the number. The structure of the program becomes simple using this strategy.

The significant thing here is that if a new task starts or if a previous task restarts, the same R program is executed. If a new task starts, it is quite natural to execute a program for such a purpose. However, if a previous task restarts, the program can take a procedure in that the process is halted and returned to the point of the previous task. In fact, this strategy can be taken to produce a series of R programs. However, if all possible records of tasks are treated, the structure of R programs inevitably becomes complicated. This difficulty can be overcome by taking the strategy in which the concept of returning a point for a requisite task is relinquished and that of activating a new R program for the task is adopted.

For example, in (8) of `double1()` described previously, the command `double2(aa)` activates `double2()` with the argument `aa`. This starts up a new task. On the other hand, in (9) of `double2()` heretofore described, `double1()` is activated. This start-up aims to input a numerical value again. That is, this is a start-up for redoing a task previously finished. This is not the procedure of suspending `double2()` and going back to `double1()`. This is just a start-up of `double1()`. Since `double1()` is activated without arguments, the following tasks by `double1()` are performed independently of the record of tasks. If the process goes back to a previous task that uses the results obtained so far, the arguments specify it. This strategy clarifies which parts of the previous tasks will be inherited.

Fig. 3.5 illustrates this concept. The progress form "Main program" to "Subroutine A" is an ordinary process. The progress form "Subroutine A" to "Subroutine B" is also an ordinary process. If the task of "Subroutine A" has to be performed again, the process does not go back to "Subroutine A", but rather "Subroutine A" is activated in "Subroutine B". Since this process is not a retrogression to a previous stage, the depth of the subroutine increases. However, the process outwardly appears to go back to "Subroutine A" from "Subroutine B" to restart the task of "Subroutine A". Finally, when the job ends in "Subroutine C", the process does not go back to "Main program", but a forced quit is executed in "Subroutine C". Even if the job ends in "Subroutine B", the program can be constructed to allow forced quit to end

Depth of subroutine

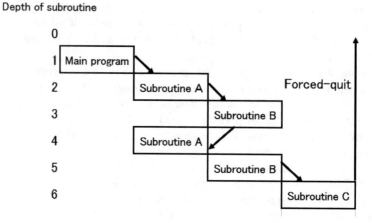

Figure 3.5   Structure of an interactive R program

the job. Forced quit in a default setting displays an error message; this may be baffling to users. To avoid such a situation, the function `exit1()` that prevents the emergence of an error message is defined in `double1()` (Program (3 - 5)) and `double2()` (Program (3 - 6)) and the function is used when forced quit is executed. That is, the depth of the subroutines increases monotonically. Hence, it does not go back to the previous depth. Forced quit is the only way to terminate the routine. This strategy, which looks rough at first glance, makes the structures of the interactive R programs simple; hence, it is easy to improve or modify completed interactive R programs. This is because each R program is directed by arguments and has a clear purpose.

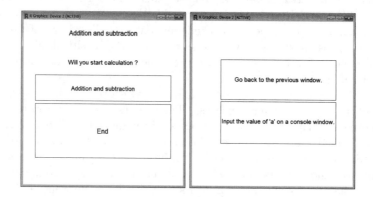

Figure 3.6   Selection of a task in the process of an interactive job given by Programs (3 - 7), (3 - 8), (3 - 9), and (3 - 10) (left). Input of the value of "a" in the process of an interactive job given by Programs (3 - 7), (3 - 8), (3 - 9), and (3 - 10) (right).

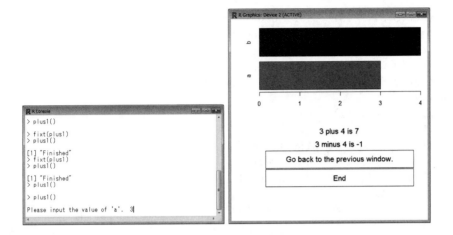

Figure 3.7   Input of the value of "a" in the process of an interactive job given by Programs (3 - 7), (3 - 8), (3 - 9), and (3 - 10) (left). Display of the results in the process of an interactive job given by Programs (3 - 7), (3 - 8), (3 - 9), and (3 - 10) (right).

Another example of R programs based on this strategy is given by Programs (3 - 7), (3 - 8), (3 - 9), and (3 - 10).

Program (3 - 7)

```
function ()
{
# The name of this R program should be plus1().
# (1)
  exit1 <- function() {
    opt <- options(show.error.messages = F)
    on.exit(options(opt))
    stop()
  }
# (2)
  par(mfrow = c(1, 1), mai = c(0.1, 0.1, 0.1, 0.1),
    omi = c(0.1, 0.1, 0.1, 0.1))
# (3)
  plot(c(0, 1), c(0, 1), xlab = "", ylab = "", type = "n",
    axes = F)
  text(0.5, 0.99, labels = "Addition and subtraction",col = 3,
    cex = 1.4)
  text(0.5, 0.99, labels = "Addition and subtraction",
    cex = 1.4)
  text(0.5, 0.78, labels = "Will you start calculation ?",
    cex = 1.3)
```

```
  rect(0.01, 0.51, 0.99, 0.69)
  text(0.5, 0.6, "Addition and subtraction", col = 1,
    cex = 1.3)
  rect(0.01, 0.11, 0.99, 0.49)
  text(0.5, 0.3, "End", cex = 1.4, col = 1)
# (4)
  p1<- locator(1)
# (5)
  if(p1$y < 0.7 && p1$y > 0.5){
    dev.off(which = dev.prev())
    dev.set(which = dev.next())
    plus2()
  }
  else{
# (6)
    dev.off(which = dev.prev())
    writeLines(sep = "\n","")
    print("Finished")
    exit1()
  }
}
```

(1) The function of `exit1()` is defined. This function deters the display of error messages when forced quit is executed.

(2) The coordinate axes in a graphics window are set.

(3) A selection on a graphics window is prompted (Fig. 3.6 (left)).

(4) The command `locator(1)` detects the position of the point where the mouse is clicked in a graphics window. The values are stored in `p1`.

(5) If `p1$y` is larger than 0.5 and less than 0.7, `dev.off(which = dev.prev())` closes the graphics window. The command `dev.set(which = dev.next())` opens a new graphics window. The function `plus2()` is executed.

(6) If `p1$y` does not satisfy the condition that it is larger than 0.5 and less than 0.7, `dev.off(which = dev.prev())` closes the graphics window. The command `writeLines(sep = "\n","")` inserts a line feed in a console window. The command `print("Finished")` displays `print"Finished"` in a console window. The command `exit1()` terminates the task. No error message is displayed.

Program (3 - 8)

```
function ()
{
# The name of this R program should be plus2().
# (1)
  bringToTop()
# (2)
  plot(c(0, 1), c(0, 1), xlab = "", ylab = "", type = "n",
```

```
  axes = F)
  text(0.5, 0.7, labels = "Go back to the previous window.",
    col = 1, cex = 1.3)
  rect(0.01, 0.51, 0.99, 0.89)
  text(0.5, 0.3, labels = "Input the value of 'a'", col = 1,
    cex = 1.3)
  rect(0.01, 0.09, 0.99, 0.49)
# (3)
  p1<- locator(1)
# (4)
  if(p1$y > 0.1 && p1$y < 0.5){
    dev.off(which = dev.prev())
    repeat{
      writeLines(sep = "\n","")
      input1 <- readline("Please input the value of 'a'.   ")
      if(input1 != "" && length(input1[is.na(charmatch
        (unlist(strsplit(input1, split = NULL)), c(as.character
        ( c(".","-",seq(from = 0, to = 9))))))]) == 0){
        aa <- as.numeric(input1)
        dev.set(which = dev.next())
        plus3(aa)
      }
      else {
        print("Please input it again.")
      }
    }
  }
# (5)
  else{
    dev.off(which = dev.prev())
    dev.set(which = dev.next())
    plus1()
  }
}
```

(1) The command **bringToTop()** displaces the working window to the front.
(2) A selection on a graphics window is prompted (Fig. 3.6 (left)).
(3) The command **locator(1)** detects the position of the point where the mouse is clicked in a graphics window. The values are stored in **p1**.
(4) If **p1$y** is larger than 0.1 and less than 0.5, the command **readline()** allows the input of the value of "a" in a console window (Fig. 3.7 (left))
(5) If **p1$y** does not satisfy the condition that it is larger than 0.1 and less than 0.5, **dev.off(which = dev.prev())** closes the graphics window. The com-

mand `dev.set(which = dev.next())` opens a new graphics window. The function `plus1()` is executed.

Program (3 - 9)

```
function (aa)
{
# The name of this R program should be plus3().
# (1)
  bringToTop()
# (2)
  plot(c(0, 1), c(0, 1), xlab = "", ylab = "", type = "n",
  axes = F)
  text(0.5, 0.7, labels = "Go back to the previous
  window.", col = 1, cex = 1.3)
  rect(0.01, 0.51, 0.99, 0.89)
  text(0.5, 0.3, labels = "Input the value of 'b'", col = 1,
  cex = 1.3)
  rect(0.01, 0.09, 0.99, 0.49)
# (3)
  p1 <- locator(1)
# (4)
  if(p1$y > 0.1 && p1$y < 0.5){
    dev.off(which = dev.prev())
    repeat{
      writeLines(sep = "\n","")
      input1 <- readline("Please input the value of 'b'.  ")
      if(input1 != "" && length(input1[is.na(charmatch
        (unlist(strsplit(input1, split = NULL)), c(as.character
        (c(".","-",seq(from = 0, to = 9))  ))))]) == 0){
        bb <- as.numeric(input1)
        dev.set(which = dev.next())
        plus4(aa, bb)
      }
      else {
        print("Please input it again.")
      }
    }
  }
  else{
# (5)
  dev.off(which = dev.prev())
  dev.set(which = dev.next())
  plus2()
  }
}
```

The value of "b" is input in the same manner as that of **plus2()**. After the input, **plus4(aa, bb)** is activated.

Program (3 - 10)

```
function(aa, bb)
{
# The name of this R program should be plus4().
# (1)
  exit1 <- function() {
    opt <- options(show.error.messages = FALSE)
    on.exit(options(opt))
    stop()
  }
# (2)
  bringToTop()
# (3)
  par(mfrow = c(2, 1), mai = c(0.8, 0.7, 0.1, 0.1),
   omi = c(0.1, 0.1, 0.1, 0.1))
  barplot(c(aa, bb), names = c("a","b"), col = c("red", "blue"),
   horiz = T)
# (4)
  plot(c(0, 1), c(0, 1), xlab = "", ylab = "", type = "n",
   axes = F)
  text1 <- paste(as.character(aa), " plus ", as.character(bb),
   " is ", as.character(aa + bb), sep = "")
  text(0.5, 0.9, labels = text1, col = 1, cex = 1.3)
  text1 <- paste(as.character(aa), " minus ", as.character(bb), "
   is ", as.character(aa - bb), sep = "")
  text(0.5, 0.7, labels = text1, col = 1, cex = 1.3)
# (5)
  text(0.5, 0.45, labels = "Go back to the previous window.",
   col = 1, cex = 1.3)
  rect(0.01, 0.31, 0.99, 0.59)
  text(0.5, 0.15, labels = "End", col = 1, cex = 1.3)
  rect(0.01, 0.01, 0.99, 0.29)
# (6)
  p1<- locator(1)
# (7)
  if(p1$y > 0.0 && p1$y < 0.3){
    dev.off(which = dev.prev())
    writeLines(sep = "\n","")
    print("Finished")
    exit1()
  }
  else{
```

```
# (8)
   dev.off(which = dev.prev())
   dev.set(which = dev.next())
#    tashizan3(aa)
   plus3(aa)
   }
}
```

(1) The function of `exit1()` is defined. This function deters the display of error messages when forced quit is executed.
(2) The command `bringToTop()` displaces the working window to the front.
(3) The command `barplot` illustrates the values of "a" and "b".
(4) The values of "a+b" and "a-b" are displayed on the graphics window (Fig. 3.7 (right)).
(5) A selection on a graphics window is prompted.
(6) The command `locator(1)` detects the position of the point where the mouse is clicked in a graphics window. The values are stored in `p1`.
(7) If `p1$y` is larger 0.0 and less than 0.3, the task is terminated.
(8) If `p1$y` does not satisfy the condition that it is larger than 0.0 and less than 0.3, `dev.off(which = dev.prev())` closes the graphics window. The command `dev.set(which = dev.next())` opens a new graphics window. The function `plus3()` is executed.

3.4   READING DATA FROM A DATA FILE

The retrieval of a file in an interactive R program is realized by the selection of a folder and a file (Fig. 3.8(top)). Then, a file name is displayed in a console window. The use of the file is requested, and if the reply is "Yes", the file will be used. This is a typical process of file retrieval. An example of the R program for this purpose is Program (3 - 11).

```
Program (3 - 11)
function ()
{
# (1)
  repeat{
# (2)
    f1 <- file.choose()
    writeLines(sep = "\n","")
    print(f1)
# (3)
    writeLines(sep = "\n","")
    input1 <- readline("Do you use this file ('1' for Yes, '2' for
    No) ?  ")
    if(input1 != "" && length(input1[is.na(charmatch(unlist
```

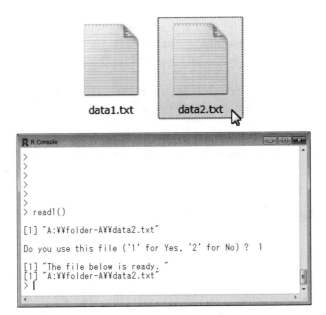

Figure 3.8 Selection of a file during retrieval of a file using Program (3 - 11) (top). Display of a file name during retrieval of a file by Program (3 - 11) (bottom).

```
    (strsplit(input1, split = NULL)), c(as.character(c(".","-",
    seq(from = 0, to = 9))))))]) == 0)
    {
      an1 <- as.numeric(input1)
    }
# (4)
    else {
      an1 <-  2
    }
# (5)
    if(an1 == 1) break
  }
# (6)
  writeLines(sep = "\n","")
  print("The file below is ready. ")
  print(f1)
}
```

(1) The command **repeat** iterates a procedure until the selection of a file ends.

(2) The command **file.choose()** allows the selection of a file (Fig. 3.8 (top)). The name of the selected file (containing the name of the folder) is

stored in **f1**. The command **writeLines(sep = "\n","")** inserts a line feed in a console window. The command **print(f1)** displays the name of the file. (3) The command **readline()** asks whether the file whose name is displayed in a console window will be used.
(4) If the input(**input1**) contains something other than numerical values, subtraction signs, and decimal points, 2 is assigned to **an1** to prompt another input.
(5) If the value of **an1** is 1, **repeat** terminates the job.
(6) The name of the selected file is displayed in a console window (Fig. 3.8 (bottom)).

3.5   MOVING DATA ON A NATURAL SPLINE

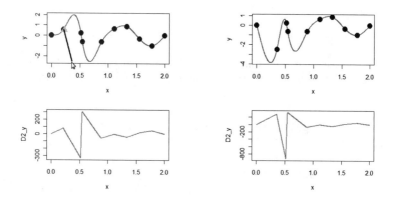

Figure 3.9   Moving a data point of the diplayed natural spline illustrated by Program (3 - 12) (left). Result of moving a data point of diplayed natural spline illustrated using Program (3 - 12) (right).

A data point in a graphics window is moved by the double-click of the mouse. The R program for this purpose is exemplified by Program (3 - 12).

```
Program (3 - 12)
function ()
{
# (1)
  nd <- 10
  xx <- seq(from = 0, to = 2, length = nd)
  yy <- sin(pi * xx * 2)
# (2)
  repeat{
```

```
# (3)
    close.screen(all.screen = T)
# (4)
    split.screen(figs = matrix(c(0.1, 0.9, 0.45, 1.0, 0.1,
      0.9, 0.0, 0.55),
      nrow = 2, byrow = T))
# (5)
    data1 <- data.frame(x = xx, y = yy)
    ex <- seq(from = min(xx), to = max(xx), by = 0.001)
    sp.pro1 <- splinefun(xx, yy, method = "natural")
    ey <- sp.pro1(ex)
    eyd2 <-  sp.pro1(ex, deriv = 2)
# (6)
    screen(2)
    plot(ex, eyd2, type = "n", xlab = "x", ylab = "D2_y")
    lines(ex, eyd2, lwd = 2, col = "magenta")
# (7)
    screen(1)
    plot(ex, ey, type = "n", xlab = "x", ylab = "y")
    lines(ex, ey, lwd = 2, col = "red")
    points(xx, yy, pch = 16, col = "blue", cex = 2)
# (8)
    del1x <- ( max(xx) - min(xx) ) * 0.03
    del1y <- ( max(ey) - min(ey) ) * 0.05
# (9)
    repeat{
      p1 <- locator(1)
      pxx <- p1$x
      pyy <- p1$y
# (10)
      for(ii in 1:nd){
        dis1x <- abs(xx[ii] - pxx)
        dis1y <- abs(yy[ii] - pyy)
        kk <- ii
        if(dis1x < del1x && dis1y < del1y ) break
      }
# (11)
      if(dis1x < del1x && dis1y < del1y ) break
    }
# (12)
    points(pxx, pyy, pch = 17, col = 5, cex = 2)
# (13)
    p1 <- locator(1)
    pxx <- p1$x
    pyy <- p1$y
```

```
# (14)
    lines(c(xx[kk], pxx), c(yy[kk], pyy), lwd = 3,
      col = "black")
    xx[kk] <- pxx
    yy[kk] <- pyy
  }
}
```

(1) The number of data (nd) is set to be 10. The values of the x-axis of the data are stored in the xx. The values of the y-axis of the data are stored in yy.

(2) The command repeat iterates the procedure of moving data and obtaining a natural spline based on the renewed data.

(3) The command close.screen(all.screen = T) deletes the setting of the graphics window.

(4) The command split.screen() divides the graphics window.

(5) xx and yy are put together in data1. The estimates given by a natural spline are set as ex. The command splinefun() produces the R program (sp.pro1()); sp.pro1() outputs estimates interpolated using a natural spline. The command sp.pro1(ex) calculates the estimates given by a natural spline at ex. The argument method = "natural" specifies the use of a natural spline. If method = is not specified, the resultant spline is constructed on the basis of different conditions. The command sp.pro1(ex, deriv = 2) outputs the second derivatives of a natural spline at ex.

(6) The command screen(2) specifies that a graph is drawn in the second area given by split.screen(). The commands plot() and lines() draw the second derivatives (eyd2).

(7) The command screen(1) specifies that the graph is drawn in the first area given by split.screen(). The interpolated values(ey) are drawn using lines. Data (yy) are plotted by points.

(8) A criterion stating the position of clicking of the mouse is close to the data is given. If the distance along the x-axis is less than del1x and that along the y-axis is less than del1y, the position of clicking is found to be close to the data. The change in the values of 0.03 and 0.05 alters the definition of closeness.

(9) The command locator(1) detects the position of the point where the mouse is clicked in a graphics window; the value of the x-axis is stored in pxx and the value of the y-axis is stored in pyy.

(10) Data are searched to find a data located in the neighborhood of the point that pxx and pyy specify. If a neighboring data is found, the number of data points is stored as kk. The command break terminates the repetition by for().

(11) If the repetition by for() is terminated in (10) as a result of finding the neighboring data, the repetition by repeat finishes. If the repetition by

`for()` is terminated in (10) because of the completion of the repetition, the repetition by `repeat` waits for another click.

(12) The command `points()` plots the points where the mouse was clicked.

(13) The command `locator(1)` detects the position of the point where the mouse is clicked in a graphics window; the value of the x-axis is stored in `pxx` and the value of the y-axis is stored in `pyy`. The data point specified previously moves to this position.

(14) The move data is illustrated by `lines()`. The values of the coordinate axes of the moved data (`pxx` and `pyy`) is stored in the position of kk-th data.

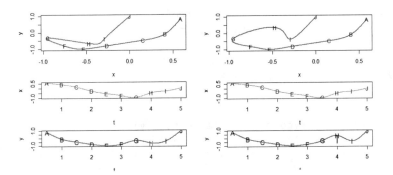

Figure 3.10   Previous positions of data (left). Updated positions of data (right).

When a curve is designed on a plane using parameters, the application of a natural spline and the moving data by the mouse allow users to draw a curve in mind. The second top graph in Fig. 3.10 (left) shows $x = x(t)$ (t is a parameter). The third top graph shows $y = y(t)$. By moving several data comprising these two functions, the curve in the top graph is altered (Fig. 3.10 (right)). Program (3 - 13) enables this work.

Program (3 - 13)

```
function ()
{
# (1)
  gys <- function(tt, yy, ey){
    screen(1)
    nd <- length(tt)
    plot(et, ey, type = "n", xlab = "t", ylab = "y")
    lines(et, ey, col = "red", lwd = 1)
    points(tt, yy, pch = LETTERS[1:nd], col = "blue", cex = 1)
  }
# (2)
  gyl <- function(tt, yy, ey){
```

```
   screen(1)
   nd <- length(tt)
   plot(et, ey, type = "n", xlab = "t", ylab = "y")
   lines(et, ey,  col = "red", lwd = 2)
   points(tt, yy, pch = LETTERS[1:nd], col = "blue", cex = 1.2)
 }
# (3)
  gxs <- function(tt, xx, ex){
   screen(2)
   nd <- length(tt)
   plot(et, ex, type = "n", xlab = "t", ylab = "x")
   lines(et, ex, lwd = 1, col = "red")
   points(tt, xx, pch = LETTERS[1:nd], col = "blue", cex = 1)
  }
# (4)
  gxl <- function(tt, xx, ex){
   screen(2)
   nd <- length(tt)
   plot(et, ex, type = "n", xlab = "t", ylab = "x")
   lines(et, ex, lwd = 2, col = "red")
   points(tt, xx, pch = LETTERS[1:nd], col = "blue", cex = 1.2)
  }
# (5)
  gxy1 <- function(xx, yy, ex, ey){
   screen(3)
   nd <- length(xx)
   plot(ex, ey, type = "n", xlab = "x", ylab = "y")
   lines(ex, ey, lwd = 2, col = "red")
   points(xx, yy, pch = LETTERS[1:nd], col = "blue", cex = 1)
  }
# (6)
  nd <- 10
  tt <-seq(from = 0.5, to = 5, length = nd)
  xx <- sin(pi * tt * 0.4)
  yy <- cos(pi * tt * 0.4)
# (7)
  repeat{
    data1 <- data.frame(t1 = tt, x= xx)
    et <- seq(from = min(tt), to = max(tt), by = 0.01)
    sp.pro1 <- splinefun(tt, xx, method = "natural")
    ex <- sp.pro1(et)
# (8)
    data2 <- data.frame(t1 = tt, y = yy)
    sp.pro2 <- splinefun(tt, yy, method = "natural")
    ey <- sp.pro2(et)
```

```
# (9)
    split.screen(figs = matrix(c(0, 1, 0.0, 0.4, 0, 1, 0.25,
        0.65, 0, 1, 0.5, 1), nrow = 3, byrow = T), erase = T)
    gxy1(xx, yy, ex, ey)
    gys(tt, yy, ey)
    gxl(tt, xx, ex)
    scr1 <- 2
# (10)
    repeat{
        p1<- locator(1)
        ptt <- p1$x
        pyy <- p1$y
        del1t <- (max(tt) - min(tt)) * 0.02
        if (scr1 == 1) {
            del1xy <- (max(ey) - min(ey)) * 0.12
        }
        else {
            del1xy <- (max(ex) - min(ex)) * 0.12
        }
# (11)
        kk <- nd + 1
        for(ii in 1:nd){
            dis1t <- abs(tt[ii] - ptt)
            if (scr1 == 1) {
                dis1xy <- abs(yy[ii] - pyy)
            }
            else {
                dis1xy <- abs(xx[ii] - pyy)
            }
            if(dis1t < del1t && dis1xy < del1xy){
                kk <- ii
                break
            }
        }
    }
# (12)
        if(kk == nd+1){
            close.screen(all.screen = T)
            split.screen(figs = matrix(c(0, 1, 0.0, 0.4, 0, 1, 0.25,
                0.65, 0, 1, 0.5, 1), nrow = 3, byrow = T), erase = T)
            gxy1(xx, yy, ex, ey)
            if (scr1 == 1) {
                scr1 <- 2
                gys(tt, yy, ey)
                gxl(tt, xx, ex)
            }
```

```
        else {
          scr1 <- 1
          gxs(tt, xx, ex)
          gyl(tt, yy, ey)
        }
      }
# (13)
      else{
        close.screen(all.screen = T)
        split.screen(figs = matrix(c(0, 1, 0.0, 0.4, 0, 1, 0.25,
          0.65, 0, 1, 0.5, 1), nrow = 3, byrow = T), erase = T)
        if (scr1 == 1) {
          gxy1(xx, yy, ex, ey)
          gxs(tt, xx, ex)
          gyl(tt, yy, ey)
          points(tt[kk], yy[kk], pch = LETTERS[kk],
            col = "green", cex = 1.2)
        }
        else{
          gxy1(xx, yy, ex, ey)
          gys(tt, yy, ey)
          gxl(tt, xx, ex)
          points(tt[kk], xx[kk], pch = LETTERS[kk],
            col = "green", cex = 1.2)
        }
# (14)
        p1<- locator(1)
        pyy <- p1$y
# (15)
        close.screen(all.screen = T)
        split.screen(figs = matrix(c(0, 1, 0.0, 0.4, 0, 1,
          0.25, 0.65, 0, 1, 0.5, 1), nrow = 3, byrow = T),
          erase = T)
        if (scr1 == 1) {
          yy2 <-  yy[kk]
          yy[kk] <- pyy
          data2 <- data.frame(t1 = tt, y = yy)
          sp.pro2 <- splinefun(tt, yy, method  = "natural")
          ey <- sp.pro2(et)
          gxy1(xx, yy, ex, ey)
          gxs(tt, xx, ex)
          gyl(tt, yy, ey)
          lines(c(tt[kk], tt[kk]), c(yy2 , yy[kk]), lwd = 2,
            col = "black")
        }
```

```
      else {
        xx2 <-   xx[kk]
        xx[kk] <- pyy
        sp.pro1 <- splinefun(tt, xx, method = "natural")
        ex <- sp.pro1(et)
        gxy1(xx, yy, ex, ey)
        gys(tt, yy, ey)
        gxl(tt, xx, ex)
        lines(c(tt[kk], tt[kk]), c(xx2 , xx[kk]), lwd = 2,
          col = "black")
      }
    }
  }
  }
}
```

(1) The function gys() for drawing the graph of $y = y(t)$ (t is a parameter) and for making the data points fixed is defined. Since screen(1) is described, the graph is drawn in the first area that split.screen() specifies.

(2) The function gyl() for drawing the graph of $y = y(t)$ (t is a parameter) and making the data points movable is defined. Since screen(1) is described, the graph is drawn in the first area.

(3) The function gxs() for drawing the graph of $x = x(t)$ (t is a parameter) and for making the data points fixed is defined. Since screen(2) is described, the graph is drawn in the second area.

(4) The function gxl() for drawing the graph of $x = x(t)$ (t is a parameter) and for making the data points movable is defined. Since screen(2) is described, the graph is drawn in the second area.

(5) The function gxy1() for drawing a graph of $(x = x(t), y = y(t))$ on a two-dimensional flat plane is defined. Since screen(3) is described, the graph is drawn in the third area.

(6) The initial values of the 10 data points are given. The values of the parameter (tt) are $\{0.5, 1, 1.5, \ldots, 5\}$. The values of the x-axis are obtained using sin(pi * tt * 0.4). The values of the y-axis are obtained using cos(pi * tt * 0.4).

(7) The command repeat iterates the job described in parentheses. Interpolation using a natural spline is carried out. The values of tt are predictor values. The values of xx are objective variable values. The estimates are stored in ex.

(8) Interpolation using a natural spline is carried out. The values of tt are the predictor values. The values of yy are objective variable values. The estimates are stored in ey.

(9) The split.screen() divides the graphics window. The function gxy1() draws $(x(t), y(t))$. The function gys() draws $y = y(t)$. The function gxl() draws $x = x(t)$. The data drawn as $x = x(t)$ becomes movable. The command

`scr1 <- 2` records the condition under which the data drawn as $x = x(t)$ is movable.

The click of the mouse on a graph of $y = y(t)$ or $x = x(t)$ is detected by `locator()`. The value of the x-axis at this point is stored as `pxx` and that of the y-axis is stored as `pyy`. If the data display in a graph of $y = y(t)$ is movable, the value of `(max(ey) - min(ey)) * 0.12` gives the region of neighborhood. If the data display in a graph of $x = x(t)$ is movable, the value of `(max(ex) - min(ex)) * 0.12` gives the region of the neighborhood.

(11) Whether the point of a click is located in the neighborhood of data points in the graph of $y = y(t)$ or $x = x(t)$ is judged. If the point of a click is located in the neighborhood of data points, the number of the data point is stored as `kk`. However, if the point of a click is not located in the neighborhood of data points, `kk <- nd + 1` is executed; the number of data plus 1 is stored as `kk`.

(12) If the point of a click is not located in the neighborhood of data points in (11), the movable graph becomes fixed and the fixed graph becomes movable. That is, a click not in the neighborhood of data points is considered the command of the interchange of the movable graph.

(13) If the point of a click is located in the neighborhood of data points in (11), the specified data points are moved in a movable graph. The argument `col = "green"` in `points()` makes the moved data green.

(14) The command `locator()` detects another click in a graphics windows. The value of y-axis of the position of the click is stored as `pyy`.

(15) The value of the y-axis of the specified data is changed to `pyy` on a movable graph. The whole graph is redrawn.

3.6 UNDERSTANDING SIMPLE REGRESSION

The characteristics of each data are grasped intuitively using a system: when the data point in a graphics window is clicked using the mouse, the characteristics of the data are displayed in a graph or shown in a console window. Such a system is realized in the case of simple regression as in Fig. 3.11 given by Program (3 - 14).

```
Program (3 - 14)
function () {
# (1)
  xx <- c(0.32, 0.44, 0.66, 0.81, 1.16, 1.57, 1.85, 2.91)
  yy <- c(4.43, 1.77, 1.12, -0.99, -1.75, -5.80, -2.60, -9.25)
  names(yy) <- c("a1", "a2", "a3", "a4", "a5", "a6", "a7", "a8")
  nd <- length(xx)
# (2)
  data1 <- data.frame(x = xx, y = yy)
  lm1 <- lm(y~x, data = data1)
# (3)
  ey1 <-lm1$fitted
```

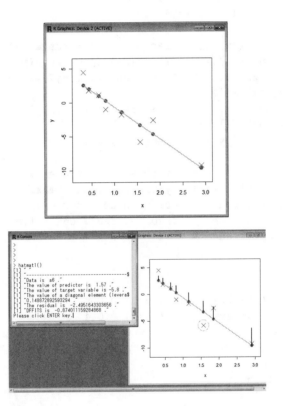

Figure 3.11   Data points and regression equation given by Program (3 - 14) (top). Characteristics of the specified data point given by Program (3 - 14) (bottom).

```
  res1 <- lm1$residuals
  df1 <- dffits(lm1)
# (4)
  xmat <- cbind(rep(1,nd), xx)
  hat1 <- xmat %*% solve(t(xmat) %*% xmat) %*% t(xmat)
# (5)
  ex <- seq(from = min(xx), to = max(xx), length = 100)
  data2 <- data.frame(x = ex)
  ey2 <- predict(lm1, data2)
# (6)
  xmin1 <- range(xx)[1] - (range(xx)[2] - range(xx)[1]) * 0.05
  xmax1 <- range(xx)[2] + (range(xx)[2] - range(xx)[1]) * 0.05
  ymin1 <- range(yy, ey2)[1] - (range(yy, ey2)[2] -
   range(yy, ey2)[1]) * 0.1
  ymax1 <- range(yy, ey2)[2] + (range(yy, ey2)[2] -
   range(yy, ey2)[1]) * 0.1
  hlen1 <- (range(yy)[2] - range(yy)[1])
# (7)
  plot(xx, yy, type = "n", xlab = "x", ylab = "y",
   xlim = c(xmin1, xmax1), ylim = c(ymin1, ymax1))
  points(xx, yy, pch = 4, col = "blue", cex = 2)
  points(xx, ey1, pch = 16, col = "red", cex = 1.5)
  lines(ex, ey2, lwd = 2, col = "green")
# (8)
  repeat{
# (9)
    del1x <- (max(xx) - min(xx)) * 0.05
    del1y <- (max(ey1) - min(ey1)) * 0.1
# (10)
    repeat{
# (11)
      p1 <- locator(1)
      pxx <- p1$x
      pyy <- p1$y
# (12)
      for(ii in 1:nd){
        dis1x <- abs(xx[ii] - pxx)
        dis1y <- abs(yy[ii] - pyy)
        kk <- ii
        if(dis1x < del1x && dis1y < del1y ) break
      }
# (13)
      if(dis1x < del1x && dis1y < del1y ) break
    }
# (14)
```

```
    plot(xx, yy, type = "n", xlab = "x", ylab = "y",
     xlim = c(xmin1, xmax1), ylim= c(ymin1, ymax1))
    points(xx, yy, pch = 4, col = "blue", cex = 2)
    points(xx, ey1, pch = 16, col = "red", cex = 1.5)
    lines(ex, ey2, lwd = 2, col = "green")
    points(xx[kk], yy[kk], pch = 1, col = "green", cex = 6)
# (15)
    for(ii in 1:nd){
       lines(c(xx[ii], xx[ii]), c(ey1[ii], ey1[ii] +
         hat1[ii, kk] * hlen1), lwd = 2)
    }
# (16)
    bringToTop(which = -1)
# (17)
    print(" ")
    print("-----------------------------------------------")
    print(paste("Data is ",names(yy[kk]), "."))
    print(paste("The value of predictor is ", xx[kk], "."))
    print(paste("The value of target variable is", yy[kk], "."))
    print("The value of a diagonal element (leverage) is ")
    print(paste( hat1[kk,kk], "."))
    print(paste("The residual is ", res1[kk], "."))
    print(paste("DFFITS is ", df1[kk], "."))
    pp <- readline("Please click ENTER key.")
  }
}
```

(1) The values of the x-axis of data is stored as **xx**. The values of the y-axis of data is stored as **yy**. The command **names()** names each element of **yy**. The number of data is stored as **nd**.

(2) **xx** and **yy** are put together in the data frame **data1**. The command **lm()** carries out a simple regression using **data1**. The result is stored in **lm1**.

(3) Estimates corresponding to data points are stored in **ey1**. Residuals are stored in **res1**. Values of DFFITS calculated using **dffits()** is stored in **df1**.

(4) A hat matrix is obtained. It is named **hat1**.

(5) The values of the x-axis of the positions for calculating estimates are named **ex**. **ex** is transformed into the data frame **data2**. The command **predict()** obtains estimates at **ex**, which are stored in **ey2**.

(6) The minimum of the region for drawing the predictor is named **xmin1**. The maximum of the same region is named **xmax1**. The minimum of the region for drawing the objective variable is named **ymin1**. The maximum of the same region is named **ymax1**. The difference between the maximum and minimum **yy** values is saved as **hlen1**.

(7) Data points and estimates are drawn in a graphics window (Fig. 3.11 (top)).

(8) The command **repeat** allows users to iterate the specification of a data point. To stop the iteration, the escape key is pressed.

(9) A criterion requiring that the position of clicking the mouse be set close to the data is given. If the distance along the x-axis is less than **del1x** and that along the y-axis is less than **del1y**, the position of clicking is found to be close to data.

(10) Data are searched to find a data point located in the neighborhood of the points that **pxx** and **pyy** specify. The command **repeat** iterates this process.

(11) The command **locator(1)** detects the position of the point where the mouse is clicked in a graphics window; the value of the x-axis is stored in **pxx** and the value of the y-axis is stored in **pyy**.

(12) Data are searched to find a data point located in the neighborhood of the points that **pxx** and **pyy** specify. If a neighboring data point is found, the number of data points is stored as **kk**. The command **break** terminates the repetition by **for()**.

(13) If the repetition by **for()** is terminated in (12) not because of completion of the repetition by **for()** but because of the finding the neighboring data, the repetition by **repeat** finishes. If the repetition by **for()** is terminated in (10) because of completion of the repetition, the repetition by **repeat** waits for another click.

(14) Data points and estimates are drawn in a graphics window. The data point specified by the mouse is plotted in green.

(15) The values of a hat matrix corresponding to the data specified by clicking the mouse are illustrated as the lengths of straight lines.

(16) The command **bringToTop()** displaces the working window to the front. The argument **which = -1** indicates a console window.

(17) Statistics on the selected data in a console window are displayed (Fig. 3.11 (bottom)).

The important features of simple regression are sometimes unveiled by observing the results of reducing some data. If some data are deleted by clicking the mouse in a console window, the effect of data reduction is grasped intuitively. Fig. 3.12, given by Program (3 - 15), shows the process for this job.

Program (3 - 15)

```
function ()
{
# (1)
  xx <- c(0.77, 0.93, 0.97, 1.17, 2.01, 2.14,
    2.19, 2.43, 3.02, 3.31, 3.42, 3.58, 3.61, 4.07)
  yy <- c(2.46, -2.31, -1.45, 2.74, 2.65, 7.51, 1.49,
    1.55, 3.70, 3.40, 2.76, 6.46, 7.96, 5.96)
# (2)
  xmin1 <- range(xx)[1] - (range(xx)[2] - range(xx)[1]) * 0.02
  xmax1 <- range(xx)[2] + (range(xx)[2] - range(xx)[1]) * 0.02
```

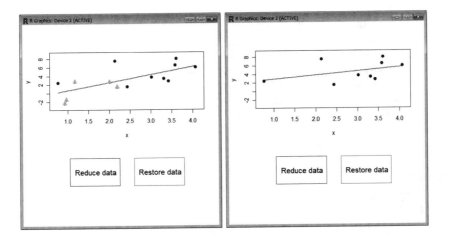

Figure 3.12   Specification of data points to be deleted by Program (3 - 15) (left).
Graph after data reduction by Program (3 - 15) (right).

```
  ymin1 <- range(yy)[1] - (range(yy)[2] - range(yy)[1]) * 0.1
  ymax1 <- range(yy)[2] + (range(yy)[2] - range(yy)[1]) * 0.1
# (3)
  xx1 <- xx
  yy1 <- yy
# (4)
  del1x <- (max(xx) - min(xx)) * 0.02
  del1y <- (max(yy) - min(yy)) * 0.05
# (5)
  repeat{
# (6)
    close.screen(all.screen = T)
    split.screen(figs = matrix(c(0, 1, 0.4, 1.0, 0, 1, 0.0,
     0.5), nrow = 2, byrow = T), erase = T)
# (7)
    screen(2)
    plot(c(0,1), c(0,1), type = "n", xlab = "", ylab = "",
     axes = F)
    rect(0.1, 0.1, 0.45, 0.9, border = "blue", lwd = 1)
    text(0.28,0.5, labels = "Reduce data", cex = 1.3)
    rect(0.55,0.1,0.9,0.9, border = "red", lwd = 1)
    text(0.73,0.5, labels = "Restore data", cex = 1.3)
# (8)
    screen(1)
    data1 <- data.frame(x = xx, y = yy)
    lm1 <- lm(y~x, data = data1)
```

```
      ex <- seq(from = min(xx), to = max(xx), length = 100)
      data2 <- data.frame(x = ex)
      ey <- predict(lm1, data2)
      plot(ex, ey, type = "n", xlab = "x", ylab = "y",
       xlim = c(xmin1, xmax1), ylim = c(ymin1, ymax1))
      lines(ex, ey, lwd = 2, col = "red")
      points(xx, yy, pch = 16, col = "blue", cex = 1.3)
# (9)
      deld <- NULL
      dnn <- 0
# (10)
      repeat{
# (11)
        repeat{
# (12)
          p1<- locator(1)
          pxx <- p1$x
          pyy <- p1$y
# (13)
          for(ii in 1:length(xx)){
            dis1x <- abs(xx[ii] - pxx)
            dis1y <- abs(yy[ii] - pyy)
            kk <- ii
            if(dis1x < del1x && dis1y < del1y) break
          }
# (14)
          if(dis1x < del1x && dis1y < del1y) break
          if(dis1x > del1x * 5 || dis1y > del1y * 5) break
        }
# (15)
        if(dis1x > del1x * 5 || dis1y > del1y * 5) break
        dnn <- dnn + 1
        deld[dnn] <- kk
        points(xx[kk], yy[kk], pch = 17, col = "green",
         cex = 1.5)
      }
# (16)
      screen(2)
      plot(c(0, 1), c(0, 1), type = "n", xlab = "", ylab = "",
       axes = F)
      rect(0.1, 0.1, 0.45, 0.9, border = "blue", lwd = 3)
      text(0.28, 0.5, labels = "Reduce data", cex = 1.3)
      rect(0.55, 0.1, 0.9, 0.9, border = "red", lwd = 3)
      text(0.73, 0.5, labels = "Restore data", cex = 1.3)
# (17)
```

```
      repeat{
        p1<- locator(1)
        pxx <- p1$x
        pyy <- p1$y
# (18)
        if(pxx > 0.1 && pxx < 0.45 ){
          sw1 <- 1
          break
        }
        if(pxx > 0.55 && pxx < 0.9 ){
          sw1 <- 2
          break
        }
      }
# (19)
      if(sw1 == 1 && dnn != 0){
        points(pxx, pyy, pch = 16, cex = 2, col = "magenta")
        xx <- xx[-deld]
        yy <- yy[-deld]
      }
      if(sw1 == 2){
        points(pxx, pyy, pch = 16, cex = 2, col = "magenta")
        xx <- xx1
        yy <- yy1
      }
      if(length(xx) <= 1) break
    }
# (20)
    if(length(xx) <= 1) {
      message("More data are needed")
      break
    }
  }
```

(1) The values of x-axis of data are stored as **xx**. The values of y-axis of data are stored as **yy**.

(2) The minimum of the region for drawing the predictor is named **xmin1**. The maximum of the same region is named **xmax1**. The minimum of the region for drawing the objective variable is named **ymin1**. The maximum of the same region is named **ymax1**.

(3) To restore the reduced data, **xx** values are stored in **xx1** and **yy** values are stored in **yy1**.

(4) A criterion requiring that the position of clicking the mouse be set close to the data is given. If the distance along the x-axis is less than **del1x** and

that along the y-axis is less than **delly**, the position of clicking is found to be close to data.

(5) The command **repeat** iterates the job of reducing data and restoring data.

(6) The command **close.screen(all.screen = T)** deletes the setting of the graphics window. The **split.screen()** divides the graphics window.

(7) Since **screen(2)** is described, the selection between "Reduce data" and "Restore data" is prompted in the second area given by **split.screen()**.

(8) Since **screen(1)** is described, the graph for showing data points and estimates are drawn in the first area given by **split.screen()**.

(9) To allow the reduction in plural data, **deld** is set to save the number of reduced data. **dnn** is set to store the number of elements (the number of reduced data) of **deld**.

(10) The command **repeat** enables reduction in plural data.

(11) The command **repeat** iterates the task until the neighboring area of data is clicked. When the neighboring area of data is clicked, **break** is executed to terminate the repetition.

(12) In a graphics window, the points where the mouse is clicked are detected by **locator()** and the value of the x-axis is saved as **pxx** and the value of the y-axis is saved as **pyy**.

(13) Data are searched to find a data point located in the neighborhood of the clicked point. If a neighboring data point is found, the number of data points is stored as **kk**. The command **break** terminates the repetition by **for()**.

(14) If the mouse clicks points in the neighborhood of a data point or the mouse clicks points far from all data points, **repeat** terminates the repetition. If both conditions are not satisfied, **repeat** indicates waiting for another click.

(15) If the mouse clicks points far from all data points, data points are not reduced and the process of selection between "Reduce data" and "Restore data" starts. For this purpose, **break** terminates the repetition by **for()**. If the mouse clicks points in the neighborhood of a data point, 1 is added to the value stored in **dnn**, the number of the specified data is stored in **deld**, and the position of data is plotted in green. Then, the process goes back to (11) to continue reduction in data.

(16) Selection between "Reduce data" and "Restore data" is prompted in the second area.

(17) The command **locator(1)** detects the position of the point where the mouse is clicked in a graphics window; the value of the x-axis is stored in **pxx** and the value of the y-axis is stored in **pyy**.

(18) If the value of **pxx** is larger than 0.1 and less than 0.45, **sw1** is set to be 1. **sw1** of 1 indicates that some data will be reduced. If the area of "Reduce data" is clicked in (12) and the same area is clicked in (17), another specification of a data point to be reduced is not carried out and the data points specified so far are deleted. That is, the area of "Reduce data" is double-clicked, and the data points specified so far are deleted. If the value of **pxx** is larger than 0.55 and less than 0.9, **sw1** is set to be 2. **sw1** of 2 indicates that the data points will be restored. If the area of "Restore data" is clicked in (12) and

the same area is clicked in (17), the data points are restored. That is, if the area of "Restore data" is double-clicked, the data points are restored. Hence, this technique provides usability as if the R program detects a double-click. (19) If `sw1` is 1, data points are deduced. If `sw1` is 2, data points are restored. (20) If the number of data is equal to or less than 1, the error message is input and the process is finished.

3.7 ADJUSTING THREE-DIMENSIONAL GRAPHS

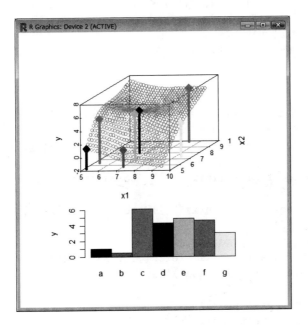

Figure 3.13   Three-dimensional display of regression surface, given by Program (3 - 16)

To grasp the behavior of the estimates given by interpolating data with two predictors using a thin-plate spline, a perspective plot of a curved surface constructed using estimates is a useful tool. If the users can grasp the responses of estimates to the alteration of the values of the target variable in sight, they can design a free-form surface efficiently. Fig. 3.13 given by Program (3 - 16) illustrates such a work.

```
Program (3 - 16)
function() {
# (1)
  library(scatterplot3d)
# (2)
```

```
  library(assist)
# (3)
  xx <- matrix(rep(0, 14), ncol = 7)
  xx[,1] <- c(5.1, 5.4); xx[,2] <- c(7.0, 5.7); xx[,3] <-
  c(8.4, 9.9)
  xx[,4] <- c(6.6, 8.1); xx[,5] <- c(7.0, 6.9); xx[,6] <-
  c(5.3, 6.4)
  xx[,7] <- c(9.3, 7.4)
  yy <- NULL
  yy[1] <- 1; yy[2] <- 2; yy[3] <- 3; yy[4] <- 4
  yy[5] <- 5; yy[6] <- 6; yy[7] <- 7
  nd <- length(yy)
  names(yy) <- letters[1:nd]
# (4)
  repeat{
# (5)
    data1 <- data.frame(x1 = xx[1,], x2 = xx[2,], y = yy)
    my.mat <- matrix(runif(25), nrow = 5)
    ssr1 <- ssr(y~x1 + x2, rk = tp.pseudo(list(x1,x2)),
     data = data1, limnla = -10)
    ex1 <- seq(from = min(xx[1,]), to = max(xx[1,]),
     by = 0.15)
    ex2 <- seq(from = min(xx[2,]), to = max(xx[2,]),
     by = 0.15)
    ex12 <- expand.grid(ex1, ex2)
    data1 <- data.frame(x1 = ex12[,1], x2 = ex12[,2])
    ey <- predict(ssr1, data1)$fit
# (6)
  zmin1 <- min(pretty(ey))
  zmax1 <- max(pretty(ey))
# (7)
  close.screen(all.screen = T)
  split.screen(figs = matrix(c(0.1, 0.9, 0.3, 1, 0.1, 0.9,
   0.0, 0.5), nrow = 2, byrow = T))
# (8)
  screen(1)
  data2 <- data.frame(x1 = xx[1,], x2 = xx[2,], y = yy)
  p3d <- scatterplot3d(data2, type = "h", lwd = 5, pch = 5,
   zlim = c(zmin1, zmax1), color = c(1, 2, 3, 4, 5, 6, 7),
   main = "")
  p3d$points3d(ex12[,1], ex12[,2], ey, col = gray(0.7))
  p3d$points3d(xx[1,], xx[2,], yy, col = c(1, 2, 3, 4, 5, 6, 7),
   type = "h", lwd = 5, pch = 5)
# (9)
  screen(2)
```

```
  barplot(yy, col = c(1, 2, 3, 4, 5, 6, 7),space = 0, xlab = "",
    ylab = "y")
# (10)
  repeat{
    p1<- locator(1)
    pxx <- p1$x
    pyy <- p1$y
    pxx <- ceiling(pxx)
    if(pxx >= 1 & pxx <= nd ) break
  }
# (11)
  yy[pxx] <- pyy
  points(pxx - 0.5, pyy, pch = 16, col = "gray", lwd = 3,
    cex = 1.3)
  }
}
```

(1) The use of the package "scatterplot3d" is described. This package is used to illustrate a perspective plot using scatterplot3d().

(2) The use of the package "assist" is described. This package is used to conduct interpolation using a thin-plate spline; ssr() is employed for this purpose.

(3) The values of the predictor of data are stored as xx. The values of the target variable of data are stored as yy. The number of data is saved as nd. The elements of yy are named letters[1:nd].

(4) The command repeat iterates the task in parentheses. This setting enables the repetition of altering the values of the target variable. Users press the escape key to stop the repetition.

(5) The command ssr() conducts interpolation using a thin-plate spline. The argument limnla = -10 carries out interpolation rather than smoothing. The estimates (interpolated values) at ex12 are saved as ey.

(6) The minimum value of the z-axis of the perspective plot is zmin1. The maximum value of the z-axis is zmax1.

(7) The command close.screen(all.screen = T) deletes the setting of the graphics window. The command split.screen() divides the graphics window.

(8) A perspective plot is illustrated in the first area. Data is put together in a data frame data2. The command scatterplot3d() draws a perspective plot using the data frame. Furthermore, scatterplot3d() with the argument col = gray(0.7) draws the curved surface produced by the estimates (interpolated values, ey) with gray. Moreover, scatterplot3d() with the argument type = "h" illustrates arrows indicating the position of the data points.

(9) The commnad barplot() draws a barplot in the second area. The argument space = 0 indicates that there is no space between bars. The argument

`space = 0` indicates that the smallest integer not less than the value of the y-axis (vertical direction) is identical to the number of bars counted from the bottom.

(10) The points where the mouse are clicked is detected by `locator()`. The value of the x-axis is stored as `pxx`. The value of the y-axis is stored as `pyy`. Since the command `ceiling(pxx)` rounds up `pxx` at the decimal point, the value of `pxx` is the number of bars counted from the bottom.

(11) The value of `pyy` is stored as `pxx`. The replaced data is plotted in gray.

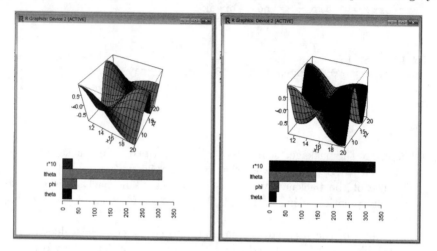

Figure 3.14   Adjustment of a curved surface with two predictors, given by Program (3 - 17).

A perspective plot is made more beneficial by locating a viewpoint and a light source approximately. Hence, if the result of changing a viewpoint and a light source of a perspective plot is viewed in a graphics window in sight, useful perspective plots are obtained easily. Fig. 3.14 given by Program (3 - 17) illustrates such a work.

Program (3 - 17)
```
function() {
# (1)
  nx1 <- 20
  nx2 <- 30
  xx1 <- seq(from = 11, to = 20, length = nx1)
  xx2 <- seq(from = 6, to = 20, length = nx1)
  xx12 <- expand.grid(xx1, xx2)
  yy <- sin(xx12[,1] * 0.13 * pi) * cos(xx12[,2] * 0.18 * pi)
  yymat1 <- matrix(yy, nrow = nx1)
# (2)
  theta1 <- 30
```

```
    phi1 <- 45
    ltheta1 <- 100
    r1 <- 300
    height1 <- c(theta1, phi1, ltheta1, r1)
# (3)
repeat{
# (4)
    close.screen(all.screen = T)
    split.screen(figs = matrix(c(0.1, 0.9, 0.25, 1.0, 0.1, 0.9,
    0.0, 0.5), nrow = 2, byrow = T))
# (5)
    persp(xx1, xx2, yymat1, theta = height1[1], phi = height1[2],
    expand = 0.9, col = "skyblue", ltheta = height1[3],
    shade = 0.75, ticktype = "detailed", xlab = "x1",
    ylab = "x2", zlab = "y", main = "",  lphi = 10,
    r = height1[4] * 0.1, d = 1,)
# (6)
    screen(2)
    names(height1) <- c("theta", "phi", "ltheta", "r*10")
    barplot(height = height1, col = c(1,2,3,4), space=0,
    horiz = T, beside = T, xlab = "", las = 2,
    xlim = c(0,360))
# (7)
    repeat{
      p1<- locator(1)
      pxx <- p1$x
      pyy <-  ceiling(p1$y)
      if(pxx >= 0 & pxx <= 360 & pyy >= 1 & pyy <= 4) break
    }
# (8)
    height1[pyy] <- pxx
    points(pxx, pyy - 0.5, pch = 16, cex = 2, col = "magenta")
    }
}
```

(1) The values of the two predictors of data are stored as **xx1** and **xx2**. The values of the target variable of data are stored as **yymat1**.

(2) The initial values of the positions of a viewpoint and a light source of a perspective plot are given. The value of **theta1** gives the azimuth of a viewpoint. That of **phi1** gives the colatitude of a viewpoint. That of **ltheta1** gives the azimuth of a light source. That of **r1** gives the distance of a light source. These four values are saved as **height1**.

(3) The command **repeat** indicates that the procedure in parentheses repeats. This process allows users to repeat the adjustment of the positions of a viewpoint and a light source of a perspective plot.

(4) The command `close.screen(all.screen = T)` deletes the setting of the graphics window. The command `split.screen()` divides the graphics window.

(5) A perspective plot is drawn in the first area. The argument `height1` given in (2) specifies the positions of a viewpoint and a light source.

(6) A bar plot is drawn in the second area. The argument `space = 0` indicates no space between bars. The argument `space = 0` indicates that the smallest integer not less than the value of the y-axis (vertical direction) is identical to the number of bars counted from the bottom.

(7) The points where the mouse is clicked is detected using `locator()`. The value of the x-axis is stored as `pxx`. The value of the y-axis is stored as `pyy`. Since the command `ceiling(pxx)` rounds up `pxx` after the decimal point, `pxx` is the number of bars counted from the bottom. If the position where the mouse is clicked is located in the area of the bar plot, the repetition terminates. Otherwise, another click is expected.

(8) The `pyy`-th value in `height1` is stored as `pxx`. The altered value is shown in magenta.

3.8   CONSTRUCTING POLYNOMIAL REGRESSION EQUATIONS INTERACTIVELY

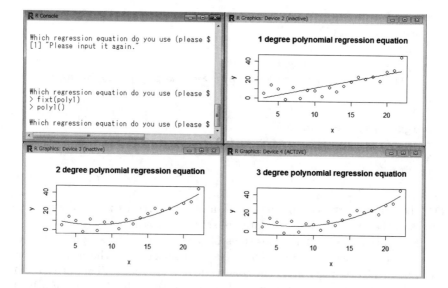

Figure 3.15   Regression to linear equation, quadratic equation, and cubic equation by Program (3 - 18).

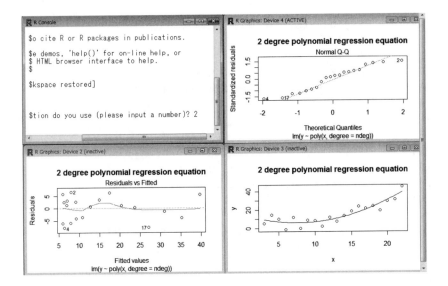

Figure 3.16   Details of quadratic regression equation given by Program (3 - 18).

When polynomial equations are regressed by the least squares method, the degree of a polynomial equation for regression is usually changed one by one to choose the most appropriate one. If the selection of the degree of a polynomial equation and the determination of the characteristics of the regression by a polynomial equation are conducted interactively and continuously, the same operation with another data is carried out easily. Fig. 3.15 and Fig. 3.16, given by Program (3 - 18), show part of such an operation.

```
Program (3 - 18)
function() {
# (1)
  reg1 <- function(xx, yy, ndeg) {
    data1 <- data.frame(x = xx, y = yy)
    lm1 <- lm(y~poly(x, degree = ndeg), data1)
    ex <- seq(from = min(xx), to = max(xx), length = 100)
    data2 <- data.frame(x = ex)
    ey <- predict(lm1, newdata = data2)
    plot(xx, yy, main = paste(as.character(ndeg),
      "degree polynomial regression equation"), xlab="x",
      ylab="y")
    lines(ex, ey)
  }
# (2)
  reg2 <- function(xx, yy, ndeg) {
    data1 <- data.frame(x = xx, y = yy)
```

```r
    lm1 <- lm(y~poly(x, degree = ndeg), data1)
    for (ii in 1:2){
      dev.set(1)
      plot.lm(lm1, which = ii, main = paste(as.character(ndeg),
       "degree polynomial regression equation"))
    }
  }
# (3)
  graphics.off()
# (4)
  xx <- c(3, 4, 5, 6, 7, 8, 9, 10, 11, 12, 13,
   14, 15, 16, 17, 18, 19, 20, 21, 22)
  yy <- c(5.16, 14.65, 10.20, -1.64, 11.82, -0.13,
   8.83, 8.33, 2.04, 11.90, 7.16, 13.49, 18.52, 23.97,
   21.64, 24.01, 19.39, 29.76, 31.26, 45.31)
# (5)
  ndegmin <- 1
  ndegmax <- 3
# (6)
  for(ndeg in ndegmin:ndegmax){
# (7)
    dev.set(1)
    reg1(xx, yy, ndeg)
  }
# (8)
  repeat{
    writeLines(sep = "\n","")
    input1 <- readline("Which regression equation do you use
     (please input a number)? ")
    if(input1 != "" && length(input1[is.na(charmatch(unlist(
     strsplit(input1, split = NULL)), c(as.character(c(".","-",
     seq(from = 0, to = 9)))))))]) == 0)
    {
# (9)
      ndeg <- as.numeric(input1)
      kk <-1
      for(ii in ndegmin:ndegmax){
        kk <- kk+1
        if(ii != ndeg) dev.off(kk)
      }
# (10)
      reg2(xx, yy, ndeg)
      break
    }
# (11)
```

```
    else {
      print("Please input it again.")
    }
  }
}
```

(1) The function `reg1()` for polynomial regression and the display of its result is defined. The values of the predictor of data are `xx`. The values of the target variable of data are `yy`. The argument `ndeg` sets the degrees of the polynomial equation. In `lm()`, `poly()` is specified to conduct polynomial regression.

(2) The function `reg2()` for illustrating a graph of polynomial regression is defined. The values of the predictor of data are `xx`. The values of the target variable of data are `yy`. The argument `ndeg` sets the degrees of the polynomial equation. In this function , if the value of `ii` is 1 or 2, `dev.set(1)` opens a new graphics window and `plot.lm()` draw graphs to show the characteristics of polynomial regression. If the value of `which =` is set to be 1 in `plot.lm()`, a graph of estimates and residuals is drawn. If the value of `which =` is set to be 2, a normal Q-Q plot for standardized residuals is drawn; "Q" stands for quantile.

(3) The command `graphics.off()` closes all graphics windows.

(4) The values of the predictor of data (`xx`) and those of the target variable (`yy`) are given.

(5) The range of the degrees of the polynomial equation is set. `ndegmin` is the minimum, `ndegmax` is the maximum.

(6) Polynomial regression is carried out using a degree in the range set in (5).

(7) The command `dev.set(1)` opens a new graphics window for each degree to execute `reg1()`.

(8) The degree of the polynomial equation is chosen in a console window.

(9) The command `dev.off(kk)` deletes all the graphics windows except the selected one.

(10) The degree (`ndeg`) is set to carry out `reg2()`.

(11) If the input in (8) is not appropriate, another input is prompted, while a warning message is displayed.

3.9 UNDERSTANDING LOCAL LINEAR REGRESSION

To understand the concept of local linear regression, users need to know about the method to obtain each estimate. An interactive program plays an important role for this purpose; it displays the outline of the regression to obtain an estimate in a graph when the estimate is specified. Fig. 3.17 given by Program (3 - 19) exemplifies it.

Program (3 - 19)
```
function()
{
```

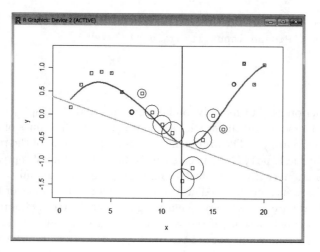

Figure 3.17   Display of implications of local linear regression given by Program (3 - 19).

```
# (1)
  lll <- function(ex, xdata, ydata, band)
  {
    x2 <- xdata - ex
    wts <- exp((-0.5 * x2^2)/band^2)
    data1 <- data.frame(x = x2, y = ydata, www = wts)
    fit.lm <- lm(y ~ x, data = data1, weights = www)
    est <- fit.lm$coef[1]
    return(est)
  }
# (2)
  bandw <- 2
# (3)
  nd <- 20
  set.seed(345)
  xx <- seq(from = 1, to = nd, by = 1)
  yy <- sin(xx * 0.4) + rnorm(nd, mean = 0, sd = 0.3)
# (4)
  ex <- seq(from = xx[1], to = xx[nd], length = 50)
  exmat <- matrix(ex, ncol = 1)
  ey <- apply(exmat, 1, lll, xdata = xx, ydata = yy,
   band = bandw)
# (5)
  xmin1 <- range(xx)[1] - (range(xx)[2] - range(xx)[1])
   * 0.05
```

```
  xmax1 <- range(xx)[2] + (range(xx)[2] - range(xx)[1])
    * 0.05
  ymin1 <- range(yy)[1] - (range(yy)[2] - range(yy)[1])
    * 0.1
  ymax1 <- range(yy)[2] + (range(yy)[2] - range(yy)[1])
    * 0.1
# (6)
  par(mai = c(1, 1, 0.5, 0.5), oma = c(0, 0, 0, 0))
  plot(xx, yy, type = "n", xlab = "x", ylab = "y",
    xlim = c(xmin1, xmax1), ylim = c(ymin1, ymax1))
  points(xx, yy, pch = 0)
  lines(ex, ey, lwd = 3, col = "red")
# (7)
  repeat {
# (8)
    p1 <- locator(1)
    pxx <- p1$x
    pyy <- p1$y
    points(pxx, pyy, cex = 2, col = "magenta", pch = 16)
# (9)
    wt <- exp(-0.5 * ((pxx - xx)/bandw)^2)
# (10)
    plot(xx, yy, type = "n", xlab = "x", ylab = "y",
      xlim = c(xmin1, xmax1), ylim = c(ymin1, ymax1))
    points(xx, yy, pch = 0)
    lines(ex, ey, lwd = 3, col = "red")
# (11)
    symbols(xx, yy, circle = sqrt(wt), inches = 0.3, add = T,
      col = "skyblue")
    abline(lsfit(xx, yy, wt), col = "green", lwd = 2)
    abline(v = pxx, col = "blue", lwd = 2)
    }
}
```

(1) The function lll() for conducting local linear regression is defined. ex(a scalar) is the value of the x-axis for estimation. xdata stores the values of the predictor of data. ydata stores the values of the target variable of data. band is the bandwidth for local linear regression. The command lm() conducts weighted simple regression using a Gaussian kernel with this bandwidth. The values in x2 are the results of subtracting ex from xdata. Hence, when simple regression using x2 is carried out, the resultant constant term is the estimate at ex. The argument weights = www gives weights. The command return(est) outputs the estimate (est), which is a constant term in the linear equation.
(2) The bandwidth (bandw) of the local linear regression is set to be 2.

(3) The number of data (nd) is given. The values of the predictor of data are
xx. The values of the target value are yy.

(4) The values of the x-axis for estimation by local linear regression are given
as ex. The transformation of ex into its matrix form yields exmat. The
command apply() executes lll() using one element of exmat as the first
argument. The second and later arguments of lll() are specified as the
arguments of apply(). The command apply() repeats the task using each
element of exmat. Hence, ey is a vector of estimates; each element of the
vector is the result of local linear regression given by one element of exmat.

(5) The minimum range of the predictor is xmin1. The maximum range is
xmax1. The minimum range of the target variable is ymin1; the maximum
range is ymax1.

(6) The data and estimates are drawn in a graph.

(7) The command repeat repeats the task of specifying the position of the
estimate.

(8) The command locator(1) detects the position of the point where the
mouse is clicked in a graphics window; the value of the x-axis is stored in pxx
and the value of the y-axis is stored in pyy. The point is plotted in magenta.

(9) The weights for calculating estimates at pxx are set to be wt.

(10) Tha data and estimates are drawn in a graph.

(11) The command symbols() draws circles at the data points; the areas of
the circles are identical to the weights (wt). The command abline() draws a
straight line that shows the linear equation used to calculate the estimate at
the specified point. The command lsfit() conducts simple regression with
data (xx, yy) and weights (wt). Furthermore, to show the position of the
x-axis used for estimation more clearly, abline() draws a vertical straight
line passing through the estimation point in blue.

To understand the bandwidth (a type of smoothing parameter) of local
linear regression, R programs such as Program (3 - 20) are useful. Program
(3 - 20) displays Fig. 3.18.

Program (3 - 20)

```
function()
{
# (1)
  lll <- function(ex, xdata, ydata, band)
  {
    x2 <- xdata - ex
    wts <- exp((-0.5 * x2^2)/band^2)
    data1 <- data.frame(x = x2, y = ydata, www = wts)
    fit.lm <- lm(y ~ x, data = data1, weights = www)
    est <- fit.lm$coef[1]
    return(est)
  }
# (2)
```

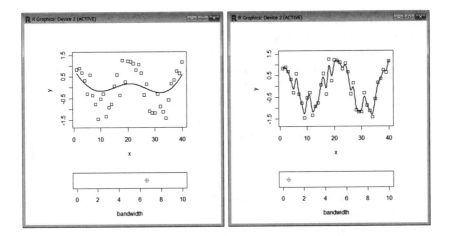

Figure 3.18   Bandwidth and regression curve of local linear regression given by Program (3 - 20).

```
nd <- 40
set.seed(121)
xx <- seq(from = 1, to = nd, by = 1)
yy <- cos(xx * 0.3) + rnorm(nd, mean = 0, sd = 0.5)
ex <- seq(from = xx[1], to = xx[nd], length = 200)
# (3)
exmat <- matrix(ex, ncol = 1)
bandw <- 1
ey <- apply(exmat, 1, lll, xdata = xx, ydata = yy,
 band = bandw)
xr <- range(xx)
yr <- range(yy, ey)
yr <- c(yr[1] - (yr[2] - yr[1]) * 0.1, yr[2]
 + (yr[2] - yr[1]) * 0.1)
# (4)
figs1 <-  matrix(c(0.1, 0.9, 0.3, 1.0, 0.1, 0.9, 0.0, 0.4),
 nrow = 2, byrow = T)
split.screen(figs = figs1)
# (5)
repeat {
# (6)
   screen(2)
   plot(bandw, 0, type = "n", xlab = "bandwidth", ylab = "",
    xlim = c(0, 10), yaxt = "n")
   points(bandw, 0, pch = 10, col = 3, cex = 1.5)
# (7)
```

```
    x0 <- locator(1)$x
    if(length(x0) < 1) break
    bandw <- x0
    bandw <- ifelse(bandw > (0.5 * (xx[nd] - xx[1]))/length(xx),
    bandw, (0.5 * (xx[nd] - xx[1]))/length(xx))
    points(bandw, 0, pch = 10, col = "red", cex = 1.5)
# (8)
    close.screen(all.screen = T)
    split.screen(figs = figs1)
# (9)
    screen(1)
    plot(xx, yy, type = "n", xlab = "x", ylab = "y", xlim = xr,
     ylim = yr, cex = 1)
    points(xx, yy, pch = 0)
    ey <- apply(exmat, 1, lll, xdata = xx, ydata = yy,
     band = bandw)
    lines(ex, ey, lwd = 2, col = "blue")
  }
}
```

(1) The function lll() for conducting local linear regression is defined.
(2) The number of data (nd) is given. The values of the predictor are saved as xx. The values of the target variable are saved as yy. The position for calculating estimates is set as ex.
(3) ex is transformed into its matrix form with one column and named exmat. The initial bandwidth (bandw) is given. The command apply() calculates all the estimates; each estimate is given by each element of ex.
(4) The matrix figs1 for giving the manner of dividing the graphics window is defined. The command split.screen() divides the graphics window using figs1.
(5) The command repeat repeats the task of calculating the estimates while changing the bandwidth. Users press the escape key to stop the repetition.
(6) The command screen(2) specifies that the graph of the bandwidth is drawn in the second area given by split.screen().
(7) Clicking the mouse in the second area inputs the bandwidth.
(8) The command close.screen(all.screen = T) deletes all displays in the graphics window. The command split.screen() divides the graphics window using figs1.
(9) The command screen(1) specifies that the data and estimates are drawn in the first area given by split.screen().

3.10 SUMMARY

1. The command repeat repeats the procedures in parentheses.

2. Users press the escape key to stop the repetition.

3. If the mouse is clicked in the graphics window, `locator()` detects the position of the mouse.

4. When the mouse is clicked, `identify()` provides an explanation at the clicked point.

5. Numbers and letters are input in a console window.

6. The commands `opt <- options(show.error.messages = F)` and `on.exit(options(opt))` deter the display of error messages when forced quit is executed.

7. `options(locatorBell = FALSE)` specifies the setting of turning off the beeping sound.

8. The command `graphics.off()` closes all graphics windows.

9. The command `graphics.off()` closes all graphics windows.

10. The command `lm()` carries out simple regression.

11. When `lm()` is conducted with the argument `weights =` that gives weights to data, a weighted simple regression is carried out. Using this technique, local linear regression is realized.

12. When `lm()` is carried out with the argument `poly()`, polynomial regression is conducted.

13. The command `plot.lm()` draws graphs for showing the characteristics of polynomial regression.

14. The command `dev.set(which = dev.next())` opens a new graphics window.

15. The command `dev.off(kk)` closes kk-th graphics window.

16. The command `bringToTop()` displaces the working window to the front.

17. The command `file.choose()` enables file selection.

18. The command `readline()` allows users to input in a console window.

19. The command `close.screen(all.screen = T)` deletes the setting of the graphics window.

20. The command `split.screen()` divides the graphics window.

21. The command `splinefun()` produces R programs for interpolation using a natural spline.

22. If `barplot()` draws a bar plot with the argument `space = 0`, no space between bars of the bar plot is specified and the smallest integer, not less than the value of the y-axis (vertical direction), is identical to the number of bars counted from the bottom.

23. The command `apply()` enables the calculation of estimates as a whole.

EXERCISES

3.1   To change the thickness of the gray color of the triangles with the value of the x-axis, a part of (6) in Program (3 - 3)(page 141) is replaced with:

```
# (6)'
  points(pxx, pyy, pch = 17, col = gray(pxx), cex = 2)
```

3.2   To change the positions of the letters, `atpen = T` is added to `identify()` in (4) of Program (3 - 4)(page 142).

3.3   To make an error message appear when the procedure ends, replace (1) of Program (3 - 6)(page 147) with:

```
# (1)'
  exit1 <- function() {
    stop()
  }
```

3.4   The combination of Programs (3 - 7)(page 151), (3 - 8)(page 152), (3 - 9)(page 154) and (3 - 10)(page 155) carry out addition and subtraction. Modify these R programs to carry out multiplication and division; see to it that the numbers cannot by divided by 0.

3.5   Program (3 - 12)(page 158) to draw estimates by simple regression while the values of data are altered interactively.

3.6   Modify Program (3 - 13)(page 161) for illustrating an old curve and a new curve together; new curves should be drawn in a different color from the previous one. The phrase "old curve" means the predecessor here. Curves older than the predecessor should not be displayed.

3.7   Modify Program (3 - 14)(page 166) to carry out a similar job with quadratic regression.

3.8   Modify Program (3 - 15)(page 170) to carry out a similar job with quadratic regression.

3.9   Modify Program (3 - 16)(page 175) to show the responses of the regression flat plate to the alterations of the values of the target variable; the flat plate is given by regression to a flat plane by the least squares method.

3.10   Modify Program (3 - 17)(page 178) to change the value of **shade** = in **persp()** interactively.

3.11   In (5) of Program (3 - 18)(page 181), three graphs for showing the characteristics of polynomial regression are displayed. Modify the program to display other graphs.

3.12   Modify Program (3 - 19)(page 183) to carry out quadratic regression for the same purpose.

3.13   Along with local linear regression, loess is also a smoothing technique using nonparametric regression. Thus, modify Program (3 - 20)(page 186) to carry out smoothing by **loess()** for the same purpose.

CHAPTER 4

GRAPHICS OBTAINED USING PACKAGES BASED ON R

4.1 INTRODUCTION

Packages enable the production of more diverse graphs. A huge variety of R-based packages for drawing graphs are available. Each package has various functions for drawing graphs. The full use of these packages will satisfy R users when they obtain graphs that perfectly suit their need. Although this chapter treats a small fraction of the various functions of each package, the author expects that descriptions in this chapter will help make graphs obtained using R-based packages clearer. Please refer to other reports, articles on the Internet, and manuals for each package when a full picture of each package is required.

The package "lattice" is not covered here. This is because the package "lattice" constructs graphs by combining diverse elements; thus, providing a concise summary is difficult. In addition, the package "lattice" is described in detail in the following books:

Deepayan Sarkar (2008): Lattice: Multivariate Data Visualization with R (Use R). Springer.

Guidebook to R Graphics Using Microsoft Windows,
First Edition. By Kunio Takezawa
Copyright © 2012 John Wiley & Sons, Inc.

Paul Murrell (2005): R Graphics. Chapman and Hall/CRC.

4.2 PACKAGE "RIMAGE"

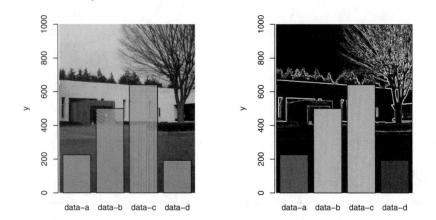

Figure 4.1   Bar chart with an image (Program (4 - 1)) (left). Bar chart with a Sobel-filtered image (right).

The package "rimage" enables (1) reading of images of a jpeg format, (2) transformation or filtering the images, and (3) drawing the resultant images.

Construction of images with graphs is also possible. Fig. 4.1 (left) given by Program (4 - 1) is an example of a graph with a jpeg format image.

```
Program (4 - 1)
function() {
# (1)
  library(rimage)
# (2)
  image1 <- read.jpeg("d:\\GraphicsR\\image2.jpg")
# (3)
  rgbm <- imagematrix(image1, type = "rgb")
# (4)
  par(omi = c(0, 0, 0, 0), mai = c(1, 2, 1, 1))
# (5)
  plot(rgbm)
# (6)
  par(new = T)
# (7)
```

```
  par(mai = c(1.25, 2, 1.25, 1))
# (8)
  yy <- c(225, 500, 641, 192)
  name1 <- c("data-a", "data-b", "data-c", "data-d")
  barplot(yy, ylab = "y", names.arg = name1, ylim = c(0, 1000),
    density = 80, angle = c(30, 60, 90, 120))
}
```

(1) The use of the package "rimage" is described.

(2) An image file (jpeg format) `image2.jpg` is retrieved and named `image1`. `image1` has an array format. It is a three-dimensional array ($389 \times 323 \times 3$) here. "389×323" indicates the number of pixels. "$\times 3$" indicates three colors: Red(R), Green(G), and Blue(B).

(3) The command `imagematrix()` constructs an image matrix and gives it the name `rgbm`. The command `imagematrix()` converts the matrix format or array format into an image matrix format. The argument `type = "rgb"` indicates that an image matrix gives a colorful image. The argument `type = "grey"` indicates that an image matrix gives a black-and-white image; `type = "gray"` does not work here.

(4) A graphics area is set.

(5) The command `plot()` gives the image of `rgbm`.

(6) The command `par(new = T)` declares that new graphs will be added, while hitherto produced graphs remain as they are.

(7) A graphics area for a bar chart is set.

(8) The command `barplot()` draws a bar chart.

The replacement of the part of (3) of Program (4 - 1) gives an image for which edge detection is applied using a Sobel filter. Fig. 4.1 (right) is the result.

```
  image2 <- sobel(image1)
  rgbm <- imagematrix(image2, type = "rgb")
```

4.3   PACKAGE "GPLOTS"

The package "gplots" enables the construction of various graphs such as graphs with plural coordinate axes, graphs with the averages, confidence intervals, and error bars, balloon plots, color samples, enhanced barplots, 2-dimensional histograms, enhanced heatmaps, graphs to show the results of lowess (a smoothing method of nonparametric regression), and Venn diagrams.

At first, let us show a graph with two or more data sets using plural coordinate axes with different scales along the y-axis. For example, Fig. 4.2 (left) is yielded using Program (4 - 2).

Program (4 - 2)
```
function(){
```

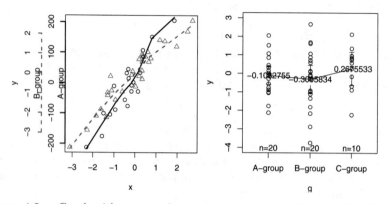

Figure 4.2   Graph with two coordinate axes along y-axis (Program (4 - 2)) (left). Averages and 95 percent confidence intervals (Program (4 - 3)) (right).

```
# (1)
  library(gplots)
# (2)
  par(mai = c(1, 1, 1, 1), omi = c(0, 0, 0, 0))
# (3)
  set.seed(111)
  xx <- rnorm(50, mean = 0, sd = 1)
  yy <- xx + rnorm(50, mean = 0, sd = 0.5)
  yy[1:20] <- yy[1:20] * 100
  gg <- c(rep("A-group", 20), rep("B-group", 30))
# (4)
  data1 <- data.frame(x = xx, y = yy, g = gg)
# (5)
  overplot(y~x|g, data = data1)
}
```

(1) The use of the package "gplots" is described.
(2) A graphics area is set.
(3) Simulation data is produced. **gg** is specified to produce the former 20 data points named the **A-group** and the latter 30 data points named the **B-group**.
(4) Data is brought together in the data frame **data1**.
(5) The command **overplot()** draws a graph with two coordinate axes along the y-axis.

The package "gplots" also draws graphs to show the averages and 95 percent confidence intervals in a graph. Fig. 4.2 (right), which is given by Program (4 - 3), is an example.

```
Program (4 - 3)
function(){
```

```
# (1)
  library(gplots)
# (2)
  par(mai = c(1, 1, 1, 1), omi = c(0, 0, 0, 0))
# (3)
  set.seed(111)
  xx <- rnorm(50, mean = 0, sd = 1)
  yy <- xx + rnorm(50, mean = 0, sd = 0.5)
  gg <- c(rep("A-group", 20), rep("B-group", 20),
    rep("C-group", 10))
# (4)
  data1 <- data.frame(x = xx, y = yy, g = gg)
# (5)
  plotmeans(y~g, data = data1, mean.labels = T, ylim = c(-4, 3))
# (6)
  points(rep(1, length = 20), yy[1:20])
  points(rep(2, length = 20), yy[21:40])
  points(rep(3, length = 10), yy[41:50])
}
```

(1) The use of the package "gplots" is described.
(2) A graphics area is set.
(3) Simulation data is produced. **gg** is specified to produce the former 20 data points named the **A-group** and the latter 30 data points named the **B-group**.
(4) Data is brought together in a data frame called **data1**.
(5) The command **plotmeans()** draws a graph to show the averages of three datasets and the 95 percent confidence intervals of the population means. The $p \cdot 100$ percent confidence interval of the population mean indicates the interval between the two values below (+ and − of ±):

$$\bar{x} \pm t((1-p) \cdot 0.5, n-1)\sqrt{\frac{v}{n}}. \tag{4.1}$$

$\bar{x}$ indicates the mean of the data. $t((1-p) \cdot 0.5, n-1)$ indicates the value that gives $((1-p) \cdot 0.5)$ as the upper probability of the t-distribution with $n-1$ degrees of freedom (n is the number of data points). v is the unbiased variance of data. p is given by specifying **p = ** as an argument of **plotmeans()**. In this example, **p = ** is not set. Hence, **p = 0.95** is assumed. For instance, the upper value of the confidence interval of the left-hand dataset in Fig. 4.2 (left) is obtained by adding the following commands to Program (4 - 3).

```
# (7)
  yy1 <- yy[1:20]
  print(mean(yy1) + sqrt(sum((yy1 - mean(yy1))^2 ) / 19 / 20)
    *   qt(0.025, df = 19, lower.tail = F))
```

sum((yy1 - mean(yy1))^2) / 19 gives the unbiased variance of yy1.

qt(0.025, df = 19, lower.tail = F) yields the t that satisfies the following equation:

$$\int_t^\infty \mathrm{pr}(x, 19)dx = 0.025. \qquad (4.2)$$

$\mathrm{pr}(x, 19)$ is a probability density function of the t-distribution with 19 degrees of freedom.

(6) Data points are added to the graph.

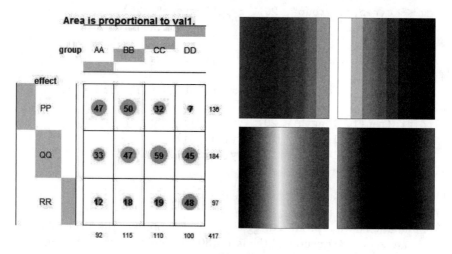

Figure 4.3 Balloon plot given by Program (4 - 4) (left). Color sample given by Program (4 - 5) (right).

The package "gplots" also produces a balloon plot. Fig. 4.3 (left), given by Program (4 - 4), exemplifies it.

Program (4 - 4)
```
function(){
# (1)
  library(gplots)
# (2)
  par(mai = c(1, 1, 1, 1), omi = c(0, 0, 0, 0))
# (3)
  gg <- c("AA", "BB", "CC", "DD")
  ee <- c("PP", "QQ", "RR")
# (4)
  set.seed(901)
  val1 <- floor(runif(12, min = 1, max = 60))
# (5)
  group <- rep(gg, 3)
  effect <- rep(ee, c(4, 4, 4))
```

```
# (6)
  balloonplot(group, effect, val1)
}
```

(1) The use of the package "gplots" is described.
(2) A graphics area is set.
(3) The first classification (four classes) is specified by **gg**. The second classification (three classes) is specified by **ee**.
(4) The number of combinations of the first and the second classifications are 12. The values of these combinations are set as **val1**.
(5) The 12 combinations are represented as three repetitions of **gg** to produce **group**. The 12 combinations are also represented as four repetitions of **ee** to produce **effect**.
(6) The command **balloonplot()** constructs a balloon plot.

The package "gplots" also yields color samples. Fig. 4.3 (right), given by Program (4 - 5), exemplifies it.

Program (4 - 5)
```
function(){
# (1)
  library(gplots)
# (2)
  par(mfrow = c(2, 2), mai = c(0.2, 0.2, 0.2, 0.2),
    omi = c(0, 0, 0, 0))
# (3)
  showpanel <- function(col){
    image(z = matrix(1:100, ncol = 1), col = col, xaxt = "n",
     yaxt = "n")
  }
# (4)
  showpanel(colorpanel(8, low = "red", high = "green"))
# (5)
  showpanel(colorpanel(8, low = "white", high = "blue"))
# (6)
  showpanel(bluered(100))
# (7)
  showpanel(greenred(100))
}
```

(1) The use of the package "gplots" is described.
(2) A graphics window is set.
(3) The function **showpanel()** for constructing a square color sample is defined.
(4) The command **colorpanel()** illustrates eight color samples from red to green on the display.
(5) The command **colorpanel()** illustrates eight color samples from white to blue on the display.

(6) The command `bluered()` illustrates 100 color samples from blue to red on the display.

(7) The command `greenred()` depicts 100 color samples from green to red on the display.

4.4 PACKAGE "GGPLOT2"

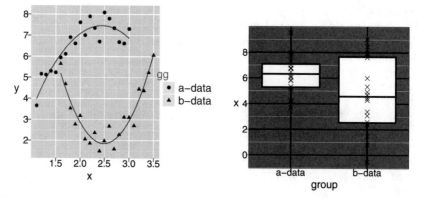

Figure 4.4   Data and regression equation given by Program (4 - 6) (left). Boxplot given by Program (4 - 7) (right).

The package "ggplot2" illustrates diverse graphs obtained using a unified methodology and aesthetic consciousness. For more details, refer to `http://had.co.nz/ggplot2/`. A glimpse is offered here. Please refer to the following text for more detail:

Hadley Wickham (2009): ggplot2: Elegant Graphics for Data Analysis (Use R). Springer.

Fig. 4.4 (left) given by Program (4 - 6) exemplifies a graph that depicts a regression to a quadratic equation using the package "ggplot2".

```
Program (4 - 6)
function() {
# (1)
  library(ggplot2)
# (2)
  set.seed(985)
  xx1 <- seq(from = 1.1, to = 3, by = 0.1)
  yy1 <- -2 * xx1 ^2  + 10 * xx1 - 5 + rnorm(20, mean = 0,
  sd = 0.5)
```

```
  gg1 <- rep("a-data", length = 20)
# (3)
  xx2 <- seq(from = 1.6, to = 3.5, by = 0.1)
  yy2 <- 5 * xx2 ^2 - 25 * xx2 + 33 + rnorm(20, mean = 0,
    sd = 0.5)
  gg2 <- rep("b-data", 20)
# (4)
  xx <- c(xx1, xx2)
  yy <- c(yy1, yy2)
  gg <- c(gg1, gg2)
# (5)
  p1 <- qplot(xx, yy, geom = c("point", "smooth"), method = "lm",
    formula = y~poly(x,2), shape = gg, xlab = "x", ylab = "y")
# (6)
  p1 + opts(axis.text.x = theme_text(size = 13), axis.text.y =
    theme_text(size = 13), axis.title.x = theme_text(size=15),
    legend.text = theme_text(size = 15,
    colour = "#0000FF"), legend.title = theme_text(size = 15,
    colour = "#FF00FF"), legend.key.size = unit(1.2, "lines"),
    plot.margin = unit(c(5, 5, 1, 1), "lines"))
}
```

(1) The use of the package "ggplot2" is described.
(2) The first dataset (**xx1, yy1**) is produced. The names of data belonging to the first dataset are stored in **gg1**.
(3) The second dataset (**xx2, yy2**) is produced. The names of data belonging to the second dataset are stored in **gg2**.
(4) **xx** and **yy** are constructed by combining the first and second datasets. **gg** is produced by combining **gg1** and **gg2**.
(5) The command **qplot()** illustrates data and the result of regression. The argument **geom** = c("point", "smooth") specifies that data points are shown by points, and the estimates obtained by regression are shown as a smooth line. The argument **method** = "lm" indicates that a linear regression is carried out. The argument **formula** = y~poly(x,2) indicates that data are regressed to a quadratic equation. The argument **shape** = gg identifies to which of the two datasets each data point belongs.
(6) The specifications of the graph constructed in (5) are added. The argument **axis.text.x** = theme_text(size = 13) sets the size of the explanation along the x-axis. The argument **axis.text.y** = theme_text(size = 13) sets the size of the explanation along the y-axis. The argument **legend.text** = theme_text(size=15, colour = "#0000FF") specifies the size and color of the letters in the legend; color = does not work here. The argument **legend. title** = theme_text(size = 15, colour = "#FF00FF") gives the size and color of the letters of the title of the legend. The argument **legend.key.size** = unit(1.2, "lines") sets the size of the legend. The argument **plot.margin**

= unit(c(5, 5, 1, 1), "lines") sets the sizes of the margins of the graph. Confidence intervals are illustrated translucently on the display. Files with a postscript format, however, do not deal with translucence. Hence, confidence intervals are not stored in such files. Furthermore, for storing graphs given by the package "ggplot2" in postscript files, **ggsave()** is executed. An example command is as follows.

```
ggsave(file = "d:\\GraphicsR\\ggplot2_1.ps", width = 5,
@height = 4.5)
```

Program (4 - 6) takes a unique approach in that it sets the specifications of graphs after each graph is constructed for use of the package "ggplot2". The package "ggplot2" takes a specific step when something is added to a graph. Fig. 4.4 (right) given by Program (4 - 7) exemplifies this.

```
Program (4 - 7)
function() {
# (1)
  library(ggplot2)
# (2)
  set.seed(986)
  xx1 <- rnorm(20, mean = 6, sd = 1.5)
  gg1 <- rep("a-data", length = 20)
  xx2 <-rnorm(20, mean = 5, sd = 3)
  gg2 <- rep("b-data", 20)
# (3)
  xx <- c(xx1, xx2)
  gg <- c(gg1, gg2)
# (4)
  p1 <- qplot(gg, xx, xlab = "group", ylab = "x")
# (5)
  p1 <- p1 + layer(geom = "boxplot", group = gg, size = 1.5)
# (6)
  p1 <- p1 + layer(geom = "point", size = 4, shape = 4)
# (7)
  p1 + opts(axis.text.x = theme_text(size = 18), axis.text.y =
    theme_text(size = 18), axis.title.x = theme_text(size = 20),
    axis.title.y = theme_text(size = 20), panel.background =
    theme_rect(fil = "#FF0012"), panel.grid.major =
    theme_line(colour = "blue", size = 1.2))
}
```

(1) The use of the package "ggplot2" is described.
(2) Simulation data is produced. **xx1** is the data of the first group; **gg1** indicates it. **xx2** is the data of the second group; **gg2** indicates it.
(3) The combination of **xx1** and **xx2** gives **xx**. The combination of **gg1** and **gg2** gives **gg**.

(4) The command `qplot()` illustrates a graph that shows the result of the classification of **xx** using **gg**.

(5) The command `layer()` indicates that something is added to the current graph. The argument **geom = "boxplot"** shows that a boxplot is drawn. The argument **group = gg** indicates the data for drawing the boxplot. The argument **size = 1.5** indicates the line thickness of the boxplot.

(6) Data points are added because the data points drawn in (4) are deleted by the boxplot constructed in (5). The argument **geom = "point"** indicates that data points are added. The argument **size = 4** shows the size of data points. The argument **shape = 4** specifies the shape of data points.

(7) The command `opts()` adds diverse specifications of the graph; **color =** does not work here.

4.5 PACKAGE "SCATTERPLOT3D"

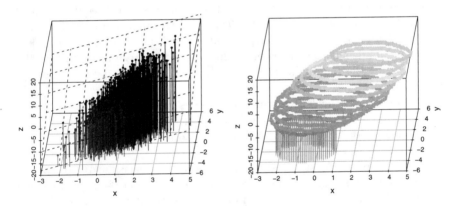

Figure 4.5   Data with two predictors and flat plane given by regression (Program (4 - 8)) (left). High-density area of data (Program (4 - 9)) (right).

The package "scatterplot3d" draws three-dimensional graphs with various forms. Since this package uses R graphics functions, the usage of letters and colors is similar to those of R. Fig. 4.5 (left) given by Program (4 - 8) exemplifies it.

```
Program (4 - 8)
function() {
# (1)
  library(scatterplot3d)
# (2)
  par(mai = c(1, 1, 1, 1), omi = c(0, 0, 0, 0), cex = 0.7)
# (3)
```

```
  set.seed(393)
  xx <- rnorm(1000, mean = 1, sd = 1)
  yy <- rnorm(1000, mean = -1, sd = 1) + xx
  zz <- runif(1000, min = -3, max = 3) + xx * 2 + yy * 2
  data1 <- data.frame(x = xx, y = yy, z = zz)
# (4)
  p3d <- scatterplot3d(data1, angle = 80, highlight.3d = T,
   type = "h", pch = 16, xlab = "x", ylab = "y", zlab = "z")
# (5)
  lm1 <- lm(z ~ x + y, data1)
# (6)
  p3d$plane3d(lm1)
}
```

(1) The use of the package "scatterplot3d" is described.
(2) A graphics area is set.
(3) Simulation data is produced for storage in the data frame `data1`.
(4) The command `scatterplot3d()` plots data points in a three-dimensional coordinate. The argument `angle = 80` specifies the angle between the x-axis and the y-axis. The argument `highlight.3d = T` indicates that the colors of data points depend on the value of the y-axis. The argument `type = "h"` indicates that the perpendicular line is dropped from the data on the x-y-plane.
(5) Regression by the least-squares method is carried out using `data1` to obtain the regresion equation $z = a_0 + a_1 x + a_2 y$.
(6) The regression flat plane yielded in (5) is added to the three-dimensional graph.

Another three-dimensional graph given by the package "scatterplot3d" is shown in Fig. 4.5 (right). This three-dimensional graph given by Program (4 - 9) depicts a high-density area of data points; the data are shown in Fig. 4.5 (left).

Program (4 - 9)
```
function() {
# (1)
  library(scatterplot3d)
# (2)
  par(mai = c(1, 1, 1, 1), omi = c(0, 0, 0, 0), cex = 0.7)
# (3)
  ffa <- function(x, y, z) {
    val1 <- 0
    dis1 <- (x - xxd)^2 +  (y - yyd)^2 + (z - zzd)^2
    s1 <- sort(dis1)
    if(s1[20] <= 5) val1 <- 1
    return(val1)
  }
```

```
# (4)
  cont1 <- function(xx, yy, vv, th1){
    nx <- length(xx)
    ny <- length(yy)
    bound1 <- NULL
    for(ii in 2:(nx-1)){
      for(jj in 2:(ny-1)){
        near1 <- c(vv[ii,jj], vv[ii-1,jj], vv[ii-1,jj], vv[ii,jj+1],
          vv[ii,jj-1])
        pp1 <- length(near1[near1 > th1])
        if(pp1 != 0  && pp1 != 5){
          bb1 <- matrix(c(xx[ii], yy[jj]), ncol = 1)
          bound1 <- cbind(bound1, bb1)
        }
      }
    }
  return(bound1)
  }
# (5)
  set.seed(393)
  xxd <<- rnorm(1000, mean = 1, sd = 1)
  yyd <<- rnorm(1000, mean = -1, sd = 1) + xxd
  zzd <<- runif(1000, min = -3, max = 3) + xxd*2 + yyd*2
  data1 <- data.frame(xxd, yyd, zzd)
# (6)
  p3d <- scatterplot3d(data1, angle = 80, type = "p",  pch = " ",
   xlab = "x", ylab = "y", zlab = "z")
# (7)
  ex <- seq(from = -3, to = 5, by = 0.1)
  ey <- seq(from = -6, to = 6, by = 0.25)
  ez <- seq(from = -20, to = 20, by = 2)
  nx <- length(ex)
  ny <- length(ey)
  nz <- length(ez)
  varray <- array(rep(0, nx * ny * nz), c(nx, ny, nz))
  for(ii in 1:nx){
    for(jj in 1:ny){
      for(kk in 1:nz){
        varray[ii, jj, kk] <- ffa(ex[ii], ey[jj], ez[kk])
      }
    }
  }
# (8)
  cola <- topo.colors(nz)
# (9)
```

```
    sw1 <- 0
    for(kk in 1:nz){
      cc1 <- cont1(ex, ey, varray[,,kk], th1 = 0.5)
      if(length(as.vector(cc1)) != 0){
# (10)
        if(sw1 == 0){
          p3d$points3d(cc1[1,], cc1[2,], rep(kk * 2-22,
            length(cc1[1,])), col = cola[kk], pch = 16, lwd = 1,
            type = "h")
          sw1 <- 1
        }
        else{
          p3d$points3d(cc1[1,], cc1[2,], rep(kk * 2 - 22,
            length(cc1[1,])), col = cola[kk], pch = 16, type = "p")
        }
      }
    }
}
```

(1) The use of the package "scatterplot3d" is described.

(2) A graphics window is set.

(3) The function `ffa()` is defined; it gives 1 to the points where data points are dense and 0 to the points where data points are sparse. The points where data points are dense are defined as the points where the 20th distance is less than or equal to 5; the distances between the points and the data points are sorted in ascending order.

(4) The function `cont1()` for selecting boundary points between high-density points and low-density ones is defined. A boundary point is defined as a point where the point and neighboring points contain both high-density and low-density points when the z of the point is fixed and the values of x and y of the points are grid point values.

(5) Simulation data is produced for storage in the data frame `data1`.

(6) The command `scatterplot3d()` draws coordinate axes of a three-dimensional graph. The arguments `type = "p"` and `pch = " "` indicates that data points are not plotted.

(7) The array `varray` is produced to give 1 to the high-density grid points and 0 to the low-density grid points in three-dimensional space. The function `ffa()` gives 1 or 0 to each element of `varray`.

(8) `cola` is set to indicate that the color of boundary points is `topo.colors(nz)` (nz is the number of grid points of the z-axis).

(9) The function `cont1()` chooses boundary points, while the z is changed one by one.

(10) Boundary points are plotted in a three-dimensional graph. When the boundary points are plotted, `type = "h"`, which indicates that a perpendic-

ular line is dropped from boundary points on the x-y-plane, is set if the z is minimal. Otherwise, type = "p" is set to plot boundary points.

4.6 PACKAGE "RGL"

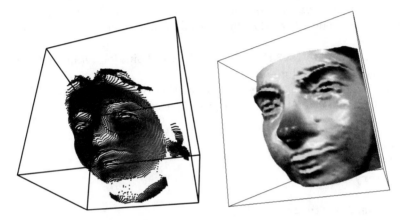

Figure 4.6 Data points in three-dimensional space (Program (4 - 10)) (left). Curved surface given by data with two predictors (Program (4 - 11)) (right).

The package "rgl" enables the production of three-dimensional graphs with various forms. This package features interactive viewpoint navigation, up to eight light sources, and rendering. For more details, refer to http://rgl.neoscientists.org/about.shtml.

Fig. 4.6 (left) given by Program (4 - 10) exemplifies a three-dimensional graph to show data points.

```
Program (4 - 10)
function() {
# (1)
  library(rgl)
# (2)
  file1 <- read.table(file = "d:\\GraphicsR\\rgl1.txt",
  header = F)
  xx <- file1[,1]
  yy <- file1[,2]
  zz <- file1[,3]
# (3)
  plot3d(xx, yy, zz, size = 1, col = "black", type = "p",
  xlab = "", ylab = "", zlab = "", axes = F)
# (4)
```

```
  view3d(theta = 15, phi = - 30)
}
```

(1) The use of the package "rgl" is described.

(2) The file rgl1.txt is retrieved and named `file1`. The values of the x-axis of the data points stored in `file1` are named `xx`. In the same way, the values of the y-axis of data points are named `yy`. The values of the z-axis of data points are named `zz`. The origin of rgl1.txt is described in the appendix.

(3) The command `plot3d()` illustrates data points in the three-dimensional graph.

(4) The command `view3d()` indicates the viewpoint for obtaining the view of the three-dimensional graph.

The package "rgl" enables the production of ordinary perspective plots. Fig. 4.6 (right) given by Program (4 - 11) exemplifies it.

Program (4 - 11)
```
function() {
# (1)
  library(rgl)
# (2)
  colfunc2 <- function(x) {
    col1 <- rev(rainbow(1000))
    colors <- col1[cut(x, breaks = length(col1),
      include.lowest = T)]
    return(colors)
  }
# (3)
  file1 <- read.table(file = "d:\\GraphicsR\\rgl2.txt",
    header = F)
  zz <- as.matrix(file1)
  xx <- seq(from = 0, to = 1, length = 100)
  yy <- seq(from = 0, to = 1, length = 150)
# (4)
  clear3d(type = "all")
# (5)
  rgl.light(theta = 50, phi = 20, viewpoint.rel = F,
    ambient = "white", diffuse = "white", specular = "white")
  rgl.light(theta = -50, phi = 20, viewpoint.rel = F,
    ambient = "white", diffuse = "white", specular = "white")
# (6)
  persp3d(xx , yy, zz, col = colfunc2(zz), xlab = "", ylab = "",
    zlab = "", axes = F)
# (7)
  user1 <- par3d("userMatrix")
  user1 <- rotate3d(user1, pi * 0.35, 1, 0, 0)
  user1 <- rotate3d(user1, -pi * 0.1, 0, 1, 0)
```

```
   user1 <- rotate3d(user1, pi * 0.05, 0, 0, 1)
# (8)
   rgl.viewpoint(theta = 60, phi = 20, userMatrix = user1)
}
```

(1) The use of the package "rgl" is described.

(2) The function `colfunc2()` for changing the color of the curved surface of the perspective plot depending on the height is defined. The height of the curved surface of the perspective plot is divided into 1000 steps. The colors of these steps are the reversed `rainbow(1000)`.

(3) The text file rgl2.txt is retrieved and named `file1`. The matrix version of `file1` is termed `zz`. The coordinates of the x-axis of grid points are stored in `xx`. Those of the y-axis are stored in `yy`. The origin of rgl2.txt is described in the appendix.

(4) The command `clear3d(type = "all")` deletes all settings concerning the three-dimensional graph.

(5) The command `rgl.light()` sets the positions and colors of lights. The maximum number of lights set by `rgl.light()` is eight. Two lights are used in this example.

(6) The command `persp3d()` depicts a perspective plot.

(7) The object is rotated. The command `par3d("userMatrix")` retrieves the default settings of the positon of an object. The setting is stored in `user1`. The command `rotate3d(user1, pi * 0.35, 1, 0, 0)` rotates `user1` around the x-axis by `pi * 0.35`. The command `rotate3d(user1, -pi * 0.1, 0, 1, 0)` rotates `user1` around the y-axis by `-pi * 0.1`. The command `rotate3d(user1, pi * 0.05, 0, 0, 1)` rotates `user1` around the z-axis by `pi * 0.05`.

(8) The command `rgl.viewpoint()` specifies the positions of the viewpoint and object. `user1` sets the position of the object.

   The package "rgl" adds three-dimensional objects to a three-dimensional graph. Fig. 4.7 (left) given by Program (4 - 12) exemplifies it.

```
Program (4 - 12)
function() {
# (1)
   library(rgl)
# (2)
   colfunc4 <- function(x) {
     col1 <- rev(heat.colors(5))
     col1 <- col1[2:5]
     colors <- col1[cut(x, breaks = length(col1),
       include.lowest = T)]
   return(colors)
}
# (3)
   fun1 <- function(xx, yy, zz){
     zz <- exp(-1.5 * (xx-2)^2 - 2 * (yy-1)^2 +
```

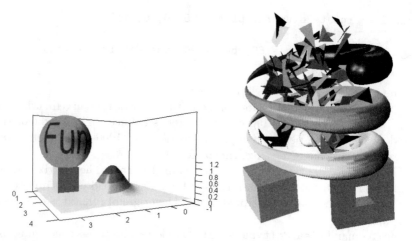

Figure 4.7   Curved surface given by data with two predictors and three-dimensional object (Program (4 - 12)) (left). Three-dimensional object (Program (4 - 13)) (right).

```
    2.5 * (xx-2) * (yy-1))
  return(zz)
  }
# (4)
  clear3d(type =  "all")
# (5)
  rgl.light( theta = 0, phi = 90, viewpoint.rel = F,
   ambient = "white", diffuse = "white", specular = "white")
  rgl.light( theta = 0, phi = -90, viewpoint.rel = F,
   ambient = "white", diffuse = "white", specular = "white")
  rgl.light( theta = 0, phi = 0, viewpoint.rel = F,
   ambient = "white", diffuse = "white", specular = "white")
  rgl.light( theta = 50, phi = 0, viewpoint.rel = F,
   ambient = "white", diffuse = "white", specular = "white")
  rgl.light( theta = 120, phi = -40, viewpoint.rel = F,
   ambient = "blue", diffuse = "blue", specular = "blue")
# (6)
  xx <- seq(from = 0, to = 4, by = 0.01)
  yy <- seq(from = -1, to = 3.5, by = 0.01)
# (7)
  zz <- outer(xx, yy, fun1)
# (8)
  rgl.surface(xx , yy, zz, col = colfunc4(zz))
# (9)
  axes3d(col = "black")
# (10)
```

```
  i1 <- c(1.0, 0.0, 0.0, 0.0)
  i2 <- c(0.0, 1.0, 0.0, 0.0)
  i3 <- c(0.0, 0.0, 1.0, 0.0)
  i4 <- c(0.0, 0.0, 0.0, 1.0)
  iden1 <- rbind(i1, i2, i3, i4)
  iden2 <- scale3d(iden1, 0.4, 0.5, 0.4)
# (11)
  g1 <- cube3d(tran = iden2, color = "skyblue")
  shade3d(translate3d(g1, 1, 0.4, 2.5))
# (12)
  rgl.spheres(1, 1.8, 2.5, radius = 1, texture =
  "d:\\GraphicsR\\rgl3eng.png")
# (13)
  user1 <- par3d("userMatrix")
  user1 <- rotate3d(user1, pi * 0.4, 1, 0, 0)
  user1 <- rotate3d(user1, -pi * 0.35, 0, 1, 0)
# (14)
  rgl.viewpoint(userMatrix = user1)
}
```

(1) The use of the package "rgl" is described.

(2) The function `colfunc4()` for changing the color of the curved surface of a perspective plot depending on the height is defined. The height of the perspective plot is divided into four steps. The colors of these steps are from the second color to the fifth one of the reversed `heat.colors(5)`.

(3) The function to be illustrated as a perspective plot is defined as `fun1()`.

(4) The command `clear3d(type = "all")` deletes all settings concerning the three-dimensional graph.

(5) The command `rgl.light()` sets the positions and colors of lights. The maximum number of lights set by `rgl.light()` is eight. Five lights are used in this example.

(6) The coordinates of the x-axis of the grid points for illustrating a perspective plot are stored in **xx**. Those of the y-axis are stored in **yy**.

(7) The command `outer()` calculates the values of the function at the grid points for illustrating a perspective plot of the function.

(8) The command `rgl.surface()` illustrates a perspective plot.

(9) The command `axes3d()` adds scales to the three-dimensional graph.

(10) A matrix **iden1** for constructing a rectangular solid is obtained. The command `scale3d()` adjusts the size of the rectangular solid. The result is stored in **iden2**.

(11) The command `cube3d()` produces **g1**. **g1** is used to construct a rectangular solid using **iden2**. The command `shade3d()` illustrates a rectangular solid using **g1**. The command `translate3d()` indicates the location of **g1**.

(12) The command `rgl.spheres()` illustrates a sphere. The first three arguments of `rgl.spheres()` indicate the position of the sphere. The argument

radius = 1 specifies the radius of the sphere. The argument `texture ="d:\\`
`GraphicsR\\rgl3eng.png"` indicates the image pasted on the sphere.
(13) The command `rotate3d()` rotates the objects.
(14) The command `rgl.viewpoint()` specifies the location of the objects.

The package "rgl" produces considerably complex three-dimensional objects. Fig. 4.7 (right) given by Program (4 - 13) exemplifies it.

Program (4 - 13)

```
function() {
# (1)
  library(rgl)
# (2)
  colfunc7 <- function(x) {
    col1 <- rev(topo.colors(100))
    colors <- col1[cut(x, breaks = length(col1),
      include.lowest = T)]
  return(colors)
}
# (3)
  clear3d(type = "all")
# (4)
  rgl.light( theta = 0, phi = 90, viewpoint.rel = F,
    ambient = "white", diffuse = "white", specular = "white")
  rgl.light( theta = 0, phi = -80, viewpoint.rel = F,
    ambient = "white", diffuse = "white", specular = "white")
  rgl.light( theta = -70, phi = -30, viewpoint.rel = F,
    ambient = "white", diffuse = "white", specular = "white")
  rgl.light( theta = 30, phi = 60, viewpoint.rel = F,
    ambient = "white", diffuse = "white", specular = "white")
# (5)
  zz <- seq(from = 0.5, to = 5.5, by = 0.01)
  yy <- sin(zz * 3) * 3
  xx <- -cos(zz * 3) * 3
# (6)
  spheres3d(xx, yy, zz, col = colfunc7(zz), radius = 0.7 *
    abs(sin(1.5 * seq(from = min(zz), to = max(zz),
    length = length(zz)))))
# (7)
  set.seed(114)
  ran1 <- runif(3000, min = -2, max = 2)
  col2 <- rainbow(100)
  kk <- 1
  for(ii in 1:100){
    v1 <- ran1[kk] + ran1[kk + 1] * 0.4
    v4 <- ran1[kk] + ran1[kk + 2] * 0.4
```

```
    v7 <- ran1[kk] + ran1[kk + 3] * 0.4
    v10 <- ran1[kk] + ran1[kk + 4] * 0.4
    v2 <- ran1[kk + 5] * 1.2 + ran1[kk + 6] * 0.4 + 0.3
    v5 <- ran1[kk + 5] * 1.2+ ran1[kk + 7] * 0.4 + 0.3
    v8 <- ran1[kk + 5] * 1.2+ ran1[kk + 8] * 0.4 + 0.3
    v11 <- ran1[kk + 5]* 1.2 + ran1[kk + 9] * 0.4 + 0.3
    v3 <- ran1[kk + 10] * 1.4 + ran1[kk + 11] * 0.4 + 4
    v6 <- ran1[kk + 10] * 1.4 + ran1[kk + 12] * 0.4 + 4
    v9 <- ran1[kk + 10] * 1.4 + ran1[kk + 13] * 0.4 + 4
    v12 <- ran1[kk + 10] * 1.4 + ran1[kk + 14] * 0.4 + 4
    kk <- kk + 15
    vert1 <- c(
    v1,  v2,  v3,  1.0,
    v4,  v5,  v6,  1.0,
    v7,  v8,  v9,  1.0,
    v10, v11, v12, 1.0)
    ind1 <- c(1, 2, 3, 4)
    shade3d(qmesh3d(vertices = vert1, indices = ind1),
      col = col2[ii])
  }
# (8)
  iden1 <- par3d("userMatrix")
  iden1 <- rotate3d(iden1, angle = -pi/8, x = 1, y = 0, z = 1)
  g1 <- oh3d(tran = iden1, color = "magenta" )
  shade3d(translate3d(g1, 2.5, 0, -1.1))
# (9)
  iden2 <- rotate3d(iden1, angle = -pi/12, x = 1, y = 0, z = 1)
  g2 <- cube3d(tran = iden2, color = "skyblue")
  shade3d(translate3d(g2, -2, 0, -1))
}
```

(1) The use of the package "rgl" is described.

(2) The command `colfunc7()` for indicating the colors of a spiral object given by combining spheres is defined. The reversed `topo.colors(100)` is used.

(3) The command `clear3d(type = "all")` deletes all settings concerning the three-dimensional graph.

(4) The command `rgl.light()` sets the positions and colors of lights. Four lights are used in this example.

(5) The coordinate axes of the positions of spheres for constructing a spiral object given by combining spheres are calculated.

(6) The command `spheres3d()` produces the sphere. The illustration of many spheres in slightly different positions constructs a tubulate object.

(7) The command `shade3d()` draws 100 quadrangles with random shapes at random positions. The argument `qmesh3d()` indicates that a quadrangle

is drawn. The argument `vertices = vert1` specifies the coordinates of the apexes of the quadrangle.

(8) The command `par3d("userMatrix")` retrieves the setting of the current coordinate axes. The command `rotate3d()` rotates the coordinate axes. The command `oh3d()` gives g1, which is used to construct a rectangular solid with a hole. The command `shade3d()` depicts a rectangular solid with a hole using g1. The command `translate3d()` sets the position of g1.

(9) The command `rotate3d()` rotates the coordinate axes set in (8). Using the coordinate axes given by the rotation, the command `cube3d()` gives g2, which is used to construct a rectangular solid. The command `shade3d()` illustrates a rectangular solid using g2. The command `translate3d()` indicates the position of g2.

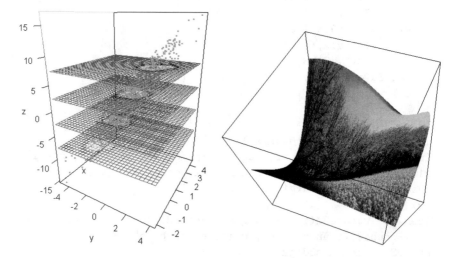

Figure 4.8   Densities of sliced three-dimensional data along z-axis (Program (4 - 14)) (left). Perspective plot in which an image is attached on the curved surface (Program (4 - 15)) (right).

The package "rgl" enables the illustration of densities of data points scattered in three-dimensional space. Fig. 4.8 (left) given by Program (4 - 14) exemplifies it.

```
Program (4 - 14)
function() {
# (1)
  library(rgl)
  library(MASS)
# (2)
  colfunc11 <- function(x) {
    col1 <- rev(rainbow(10))
```

```
    colors <- col1[cut(x, breaks = length(col1),
      include.lowest = T)]
    return(colors)
}
# (3)
    clear3d(type = "all")
# (4)
    rgl.light(theta = 0, phi = 90, viewpoint.rel = F,
      ambient = "white", diffuse = "white", specular = "white")
    rgl.light(theta = 0, phi = -90, viewpoint.rel = F,
      ambient = "white", diffuse = "white", specular = "white")
    rgl.light(theta = 0, phi = 0, viewpoint.rel = F,
      ambient = "white", diffuse = "white", specular = "white")
    rgl.light(theta = 50, phi = 0, viewpoint.rel = F,
      ambient. = "white", diffuse = "white", specular = "white")
    rgl.light(theta = 120, phi = -40, viewpoint.rel = F,
      ambient = "blue", diffuse = "blue", specular = "blue")
# (5)
    set.seed(393)
    xx <- rnorm(1000, mean = 1, sd = 1)
    yy <- rnorm(1000, mean = -1, sd = 1) + xx
    zz <- runif(1000, min = -3, max = 3) + xx*2 + yy*2
# (6)
    plot3d(xx, zz, yy, size = 3, col = "green", type = "p",
      xlab = "x", ylab = "z", zlab = "y", xlim = range(xx),
      ylim = range(yy), axes  = F, box = F)
# (7)
    xxa <- xx[zz >= 5]
    yya <- yy[zz >= 5]
    dena <- kde2d(xxa, yya, n = 30, h = c(width.SJ(xx),
      width.SJ(yy)), lims = c(range(xx), range(yy)))
    ex <- dena$x
    ey <- dena$y
    ez <- dena$z
    rgl.surface(ex ,ey, ez * 0.001 + 7.5, lwd = 1,
      col = colfunc11(ez), back = "line", front = "lines")
# (8)
    xxb <- xx[zz >= 0 & zz < 5]
    yyb <- yy[zz >= 0 & zz < 5]
    denb <- kde2d(xxb, yyb, n = 30, h = c(width.SJ(xx),
      width.SJ(yy)), lims = c(range(xx), range(yy)))
    ex <- denb$x
    ey <- denb$y
    ez <- denb$z
    rgl.surface(ex ,ey, ez * 0.001 + 2.5, lwd = 1,
```

```
  col = colfunc11(ez), back = "lines", front = "lines",
  axes = F, box = F, xaxis = F, yaxis = F, zaxis = F)
# (9)
  xxc <- xx[zz >= -5 & zz < 0]
  yyc <- yy[zz >= -5 & zz < 0]
  denc <- kde2d(xxc, yyc, n=30, h = c(width.SJ(xx),
  width.SJ(yy)), lims = c(range(xx), range(yy)))
  ex <- denc$x
  ey <- denc$y
  ez <- denc$z
  rgl.surface(ex ,ey, ez * 0.001 - 2.5, lwd = 1,
  col = colfunc11(ez), back = "lines", front = "lines",
  axes = F, box = F, xaxis = F, yaxis = F, zaxis = F)
# (10)
  xxd <- xx[ zz < -5]
  yyd <- yy[ zz < -5]
  dend <- kde2d(xxd, yyd, n = 30,h = c(width.SJ(xx),
  width.SJ(yy)), lims = c(range(xx), range(yy)))
  ex <- dend$x
  ey <- dend$y
  ez <- dend$z
  rgl.surface(ex ,ey, ez * 0.001 - 7.5, lwd = 1,
  col = colfunc11(ez), back = "lines", front = "lines",
  axes = F, box = F, xaxis = F, yaxis = F, zaxis = F)
# (11)
  axes3d(col = "black", labels = c("x", "y", "z"))
# (12)
  aspect3d(x = 1, y = 1.5, z = 1)
  view3d(theta = -60, phi = 20)
}
```

(1) The use of the package "rgl" is described. The use of the package "MASS" is described. The use of the package "MASS" is needed to use kde2d(); kde2d() derives frequencies of data of two variables.
(2) The command colfunc11() for specifying colors of a contour is defined.
(3) The command clear3d(type = "all") deletes all settings concerning the three-dimensional graph.
(4) The command rgl.light() sets the positions and colors of lights. Five lights are used in this example.
(5) Simulation data is produced.
(6) The command plot3d() plots the data points of the simulation data produced in (5) in the three-dimensional graph.
(7) The command kde2d() derives the frequencies of a histogram of the selected data; the value of the z-axis of the data is more than or equal to 5 and the variables of this data are the values of the x-axis and y-axis. The com-

mand `rgl.surface()` illustrates a perspective plot of the histogram using the frequencies. The frequencies of the z-axis values are transformed using the equation `ez * 0.001 + 7.5`. Hence, a perspective plot illustrated around the place where the corresponding data exists is substantially a contour plot.

(8) The command `kde2d()` derives the frequencies of a histogram of the selected data; the value of the z-axis of the data is less than 5 or more than or equal to 0, and the variables of this data are the values of the x-axis and y-axis. The command `rgl.surface()` depicts a perspective plot of the histogram using the frequencies. The frequencies of the z-axis values are transformed using the equation `ez * 0.001 + 2.5`. Hence, a perspective plot illustrated around a place where the corresponding data exists is substantially a contour plot.

(9) The command `kde2d()` derives the frequencies of a histogram of the selected data; the value of the z-axis of the data is less than 0 or more than or equal to −5, and the variables of this data are the values of the x-axis and y-axis. The command `rgl.surface()` illustrates a perspective plot of the histogram using the frequencies. The frequencies of the z-axis values are transformed using the equation `ez * 0.001 - 2.5`. Hence, a perspective plot depicted around a place where the corresponding data exists is substantially a contour plot.

(10) The command `kde2d()` derives the frequencies of a histogram of the selected data; the value of the z-axis of the data is less than −5 and the variables of this data are the values of the x-axis and y-axis. The command `rgl.surface()` illustrates a perspective plot of the histogram using the frequencies. The frequencies of the z-axis values are transformed using the equation `ez * 0.001 - 7.5`. Hence, a perspective plot illustrated around a place where the corresponding data exists is substantially a contour plot.

(11) The command `axes3d()` draws the coordinates axes.

(12) The command `aspect3d()` adjusts the lengths of the axes of the three-dimensional graph. The command `view3d()` specifies the point of the viewpoint.

The package "rgl" enables the attachment of an image on the curved surface of the perspective plot. Fig. 4.8 (right) given by Program (4 - 15) exemplifies it.

Program (4 - 15)

```
function() {
# (1)
  library(rgl)
# (2)
  fun1 <- function(x, y){
    zz <- sin(x) * cos(y)
  return(zz)
  }
# (3)
  clear3d(type = "all")
```

```
# (4)
  rgl.light( theta = 0, phi = -200, viewpoint.rel = F,
    ambient = "white", diffuse = "white", specular = "white")
  rgl.light( theta = 60, phi = -180, viewpoint.rel = F,
    ambient = "white", diffuse = "white", specular = "white")
# (5)
  nx <- 110
  ny <- 100
  xx <- seq(from = 1.1, to = 4, length = nx)
  yy <- seq(from = 4.25, to = 8, length = ny)
# (6)
  zz <- outer(xx, yy, fun1)
# (7)
  persp3d(xx, yy, zz, texture = "d:\\GraphicsR\\rgl6.png",
    col = "transparent", xlab = "", ylab = "", zlab = "",
    axes = F)
# (8)
  user1 <- par3d("userMatrix")
  user1 <- rotate3d(user1, pi * 1.2, 1, 0, 0)
  user1 <- rotate3d(user1, -pi * 0.01, 0, 1, 0)
  user1 <- rotate3d(user1, pi * 0.1, 0, 0, 1)
  rgl.viewpoint(userMatrix = user1)
}
```

(1) The use of the package "rgl" is described.
(2) The function `fun1()` for indicating the shape of the perspective plot is defined.
(3) The command `clear3d(type = "all")` deletes all settings concerning the three-dimensional graph.
(4) The command `rgl.light()` sets the positions and colors of lights. Two lights are used in this example.
(5) The values of the x-axis used in `func1()` are stored in `xx`. Those of the y-axis are stored in `yy`.
(6) The values given by `func1()` are stored in `zz`.
(7) The command `persp3d()` depicts a perspective plot. The argument `texture = "d:\\GraphicsR\\rgl6.png"` indicates an image, which is attached to the curved surface of the perspective plot.
(8) The perspective plot is rotated.

When many data points and a regression line are shown in a graph, the distribution data is not illustrated clearly around the place where the density of data is high. Fig. 4.9 (left) exemplifies it. A graph such as that in Fig. 4.9 (left) given by package "rgl" solves this problem. Program (4 - 16) leads to Fig. 4.9 (left).

```
Program (4 - 16)
function() {
```

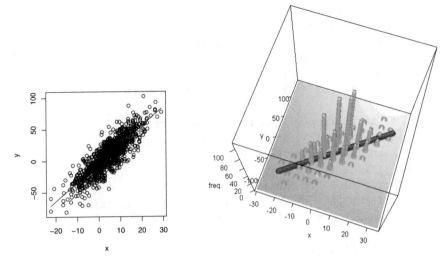

Figure 4.9   Graph that fails to show the distribution of data clearly because of too many data points (left). Three-dimensional graph given by Program (4 - 16) to show the densities of data using the heights of bars and a regression line (right).

```
# (1)
  library(rgl)
# (2)
  clear3d(type = "all")
# (3)
  rgl.light( theta = 0, phi = 30, viewpoint.rel = F,
   ambient = "white", diffuse = "white", specular = "white")
  rgl.light( theta = 0, phi = -30, viewpoint.rel = F,
   ambient = "white", diffuse = "white", specular = "white")
  rgl.light( theta = -85, phi = -80, viewpoint.rel = F,
   ambient = "white", diffuse = "white", specular = "white")
# (4)
  set.seed(435)
  nd <- 1000
  xx <- rnorm(nd, mean = 4, sd = 8)
  yy <- 3 * xx - 4 +  rnorm(nd, mean = 0, sd = 15)
  data1 <- data.frame(x = xx, y = yy)
# (5)
  lm1 <- lm(y~x, data = data1)
  ex  <- seq(from = min(xx), to = max(xx), length = 100)
  data2 <- data.frame(x = ex)
  ey <- predict(lm1, data2)
# (6)
```

```
  br1 <- pretty(n = 10 , xx)
  dx1 <- br1[2] -  br1[1]
  br1 <- c(min(br1) - dx1, br1, max(br1) + dx1)
  br2 <- pretty(n = 10, yy)
  dx2 <- br2[2] -  br2[1]
  br2 <- c(min(br2) - dx2, br2, max(br2) + dx2)
# (7)
  his2 <- matrix(rep(0, length = (length(br1) - 1) *
   (length(br2) - 1)), ncol = length(br2) - 1)
  for(kk in 1:length(xx)){
    for(jj in 1:(length(br2) - 1)){
      for(ii in 1:(length(br1) - 1)){
        if((br1[ii] <= xx[kk])&(br1[ii + 1] > xx[kk]) &
        (br2[jj] <= yy[kk]) & (br2[jj+1] > yy[kk]))
        {
          his2[ii,jj] <- his2[ii,jj] + 1
        }
      }
    }
  }
# (8)
  finep1 <- seq(from = min(br1), to = max(br1),
   length = dim(his2)[1] * 10)
  finep2 <- seq(from = min(br2), to = max(br2),
   length = dim(his2)[2] * 10)
  fineh2d <- matrix(rep(0, length = length(finep1) *
   length(finep2)), ncol = length(finep2))
  for(ii in 1:length(finep1)){
    for(jj in 3:(length(finep2)-2)){
      fineh2d[ii,jj] <- his2[(ii+9) %/% 10,(jj+9) %/% 10]
      if((ii+9) %% 10 == 0)   fineh2d[ii,jj] <- 0
      if((ii+9) %% 10 == 1)   fineh2d[ii,jj] <- 0
      if((ii+9) %% 10 == 2)   fineh2d[ii,jj] <- 0
      if((ii+9) %% 10 == 7)   fineh2d[ii,jj] <- 0
      if((ii+9) %% 10 == 8)   fineh2d[ii,jj] <- 0
      if((ii+9) %% 10 == 9)   fineh2d[ii,jj] <- 0
      if((jj+9) %% 10 == 0)   fineh2d[ii,jj] <- 0
      if((jj+9) %% 10 == 1)   fineh2d[ii,jj] <- 0
      if((jj+9) %% 10 == 2)   fineh2d[ii,jj] <- 0
      if((jj+9) %% 10 == 7)   fineh2d[ii,jj] <- 0
      if((jj+9) %% 10 == 8)   fineh2d[ii,jj] <- 0
      if((jj+9) %% 10 == 9)   fineh2d[ii,jj] <- 0
    }
  }
# (9)
```

```
  persp3d(finep1, finep2, fineh2d, axes = F,
    col = "green", xlab = "x", ylab = "y", zlab = "freq.")
# (10)
  spheres3d(ex, ey, rep(2, length = length(ex)), col = "red",
    radius = 2)
# (11)
  axes3d(col = "black",  edges = c("x--","y--","z--"))
# (12)
  aspect3d(x = 0.5, y = 0.5, z = 0.5 )
# (13)
  user1 <- par3d("userMatrix")
  user1 <- rotate3d(user1, pi * 0.26, 1, 0, 0)
  user1 <- rotate3d(user1, pi * 0.01, 0, 1, 0)
  user1 <- rotate3d(user1, -pi * 0.1, 0, 0, 1)
  rgl.viewpoint(userMatrix = user1)
}
```

(1) The use of the package "rgl" is described.
(2) The command `clear3d(type = "all")` deletes all settings concerning the three-dimensional graph.
(3) The command `rgl.light()` sets the positions and colors of lights. Three lights are used in this example.
(4) Simulation data is produced for storage in the data frame `data1`.
(5) The command `lm()` carries out simple regression. Estimates by the simple regression are stored in `ey`.
(6) The break points of the histogram for two variables are calculated and stored in `br1` and `br2`.
(7) The frequencies of the histogram for two variables are obtained using `br1` and `br2` and stored in `his2`.
(8) To depict the histogram for two variables using `his2`, the break points are segmentized to obtain `fineh2d`. The frequencies near the fringes of the bars are set to 0 to make the bars thin.
(9) The command `persp3d()` illustrates a perspective plot of `fineh2d`.
(10) The command `spheres3d()` draws the estimates by simple regression.
(11) The `axes3d()` adds coordinates axes to the three-dimensional graph. The argument `edges =` sets the position of the coordinates axes.
(12) The command `aspect3d()` adjusts the length of the axes of the three-dimensional graph.
(13) The command `rgl.viewpoint()` rotates the three-dimensional graph using `user1`.

4.7 PACKAGE "MISC3D"

The package "misc3d" enables drawing a three-dimensional contour plot, drawing points on a three-dimensional grid, drawing a three-dimensional para-

Figure 4.10   A three-dimensional object produced by Program (4 - 17); this object is constructed using parameters (left). A three-dimensional object produced by Program (4 - 18); this object is constructed by one function with three predictors (right).

metric plot, computing three-dimensional kernel density estimates, drawing interactive image slices, rendering, etc.

Firstly, this package draws three-dimensional objects represented by parameters. Fig. 4.10 (left), given by Program (4 - 17), is an example.

Program (4 - 17)
```
function() {
# (1)
  library(misc3d)
  library(rgl)
# (2)
  fx1 <- function(u, v){
    val1 <- sin(2 * u)
  return(val1)
  }
# (3)
  fy1 <- function(u, v){
    val1 <- sin(u + v * 2)
  return(val1)
  }
# (4)
  fz1 <- function(u, v){
    val1 <- cos(2 * v)
  return(val1)
  }
# (5)
```

```
  clear3d(type = "all")
# (6)
  rgl.light(theta = 75, phi = 10, viewpoint.rel = F,
    ambient = "white", diffuse = "white", specular = "white")
  rgl.light(theta = -80, phi = -20, viewpoint.rel = F,
    ambient = "white", diffuse = "white", specular = "white")
# (7)
  parametric3d(fx = fx1, fy = fy1, fz = fz1, color = "skyblue",
    umin = -pi, umax = pi, vmin = -pi, vmax = pi, n = 50)
# (8)
  user1 <- par3d("userMatrix")
  user1 <- rotate3d(user1, pi * 0.9, 1, 0 ,0)
  user1 <- rotate3d(user1, pi * 0.3, 0, 0, 1)
  user1 <- rotate3d(user1, pi * 0.1, 0, 1, 0)
  rgl.viewpoint(userMatrix = user1)
}
```

(1) The use of the packages "misc3d" and "rgl" is described. Drawing by `parametric3d()` needs the package "rgl" as well as "misc3d".
(2) The function `fx1()` for giving the values of the x-axis is defined. u and v are the parameters.
(3) The function `fy1()` for giving the values of the y-axis is defined.
(4) The function `fz1()` for giving the values of the z-axis is defined.
(5) The command `clear3d(type = "all")` deletes all settings of three-dimensional graphs.
(6) The command `rgl.light()` specifies the light settings. Two lights are set in this example.
(7) The command `parametric3d()` draws a three-dimensional object.
(8) The command `rgl.viewpoint()` rotates the three-dimensional graph using `user1`.

Furthermore, the package "misc3d" constructs a constant-height surface of a function with predictors. For example, Fig. 4.10 (right), given by Program (4 - 18), is produced.

Program (4 - 18)
```
function() {
# (1)
  library(misc3d)
# (2)
  par(mai = c(0, 0, 0, 0), omi = c(0, 0, 0, 0))
# (3)
  ff <- function(x, y, z) {
    val1 <- exp(-((x + 1)^2 + y^2 + z^2)) * 0.2 +
      exp(-((x - 1)^2 + y^2 + z^2)) * 0.3
    return(val1)
  }
```

```
# (4)
  th1 <- seq(from = 0.05, to = 0.20, length = 4)
# (5)
  cola <- rainbow(length(th1))
# (6)
  xx <- seq(from = -3, to = 2, length = 15)
  yy <- seq(from = -2, to = 2, length = 20)
  zz <- seq(from = -2, to = 2, length = 20)
# (7)
  mask1 <- function(xx, yy, zz) xx > -0.15 | zz < 0
# (8)
  screen1 <- list(z = 20, x = -70, y = 20)
# (9)
  col2 <- "white"
# (10)
  contour3d(ff, th1, xx, yy, zz, color = cola, mask = mask1,
    engine = "standard", scale = FALSE, screen = screen1,
    color2 = col2)
}
```

(1) The use of the package "misc3d" is described.
(2) A graphics area is set.
(3) The function ff() for drawing a constant-height surface is defined.
(4) th1, a threshold for drawing a constant-height surface of the function defined in (3) is given.
(5) The colors of the constant-height surface are specified in cola.
(6) The values of grid points where the values of the function ff() are calculated are given.
(7) The range in which the constant-height surface is drawn is specified in mask1. A constant-height surface is drawn in the range in which xx > -0.15 or zz < 0 is satisfied.
(8) The degrees of the rotations of the object around the axes are specified as screen1; the object is constructed using a constant-height surface.
(9) The back-side colors of the constant-height surface are set as col2.
(10) The command contour3d() draws a constant-height surface.

The command contour3d() draws an object that shows a high-density region of data points. When Fig. 4.11 (left) is viewed with the left eye and Fig. 4.11 (right) is viewed with the right eye, the object is viewed stereoscopically. Fig. 4.11 (left) is given by Program (4 - 19).

```
Program (4 - 19)
function() {
# (1)
  library(misc3d)
# (2)
  par(mai = c(0, 0, 0, 0), omi = c(0, 0, 0, 0))
```

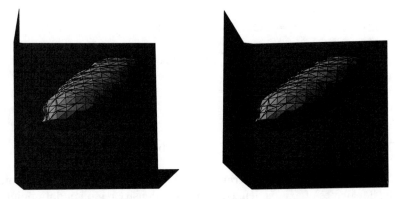

Figure 4.11   High density region of data points. A three-dimensional graph is given by Program (4 - 19) (left). Three-dimensional graph from a viewpoint slightly different from that in Fig. 4.11 (right).

```
# (3)
  ffa <- function(x, y, z) {
    nn <- length(x)
    val1 <- NULL
    for(ii in 1:nn){
      dis1 <- (x[ii] - xxd)^2 +  (y[ii] - yyd)^2 +
      (z[ii] - zzd)^2
      s1 <- sort(dis1)
      val1[ii] <- - 1/s1[20]
      if(x[ii] == -3) val1[ii] <- 1
      if(y[ii] == 6) val1[ii] <- 1
      if(z[ii] == -20) val1[ii] <- 1
    }
    return(val1)
  }
# (4)
  set.seed(393)
  xxd <<- rnorm(1000, mean = 1, sd = 1)
  yyd <<- rnorm(1000, mean = -1, sd = 1) + xxd
  zzd <<- runif(1000, min = -3, max = 3) + xxd*2 + yyd*2
# (5)
  th1 <- c(-0.4, 0.999)
  cola <- c("green", "skyblue")
# (6)
  xx <- seq(from = -3, to = 5, by = 0.5)
  yy <- seq(from = -6, to = 6, by = 0.5)
```

```
  zz <- seq(from = -20, to = 20, by = 2)
# (7)
  screen1 <- list(x = 260, y = 5, z = 0)
# (8)
  contour3d(ffa, th1, xx, yy, zz, color = cola,
    engine = "standard", scale = T, screen = screen1,
    color2 = "pink", col.mesh = "black", distance = 0.05,
    light = c(-40, 40, 1))
}
```

(1) The use of the package "misc3d" is described.

(2) A graphics area is set.

(3) The function `ffa()` for illustrating the high-density region of data points and the flat planes of axes is defined. The high-density region of data points is defined as the area in which, when squared distances between the data points and other data points are arranged in an ascending order, the inverse number of the 20th value is large. The flat planes of axes are composed of a flat plane where the value of the x-axis is -3, a flat plane where the value of the y-axis is 6, and a flat plane where the value of the z-axis is -20.

(4) The simulation data, `xxd`, `yyd`, and `zzd` are produced. Since `xxd`, `yyd`, and `zzd` are used in the function `ffa()`, `<<-` is employed.

(5) The values of the function `ffa()` on the constant-height surface are specified as `th1`. The back-side colors of the constant-height surface are set as `cola`. Although `"green"` is specified, this color is not displayed because this is an inside color of the curved surface.

(6) The grid values given by `xx`, `yy`, and `zz` are given; the value of the function `ffa()` is calculated at these points.

(7) The degrees of the rotations of the object around the axes are specified as `screen1`. When Fig. 4.11 (right) is drawn, this part is replaced with:

```
  screen1 <- list(x = 260, y = -5, z = 1)
```

(8) The command `contour3d()` draws a constant-height surface.

   When `engine = "rgl"` is specified in `contour3d()`, several functions of the package "rgl" are available. For instance, images in which objects are rotatable by a mouse are constructed. Coordinate axes can be added to a three-dimensional graph. Fig. 4.12 (left), given by Program (4 - 20), exemplifies it.

```
Program (4 - 20)
function() {
# (1)
  library(misc3d)
  library(rgl)
# (2)
  ffb <- function(x, y, z) {
    nn <- length(x)
```

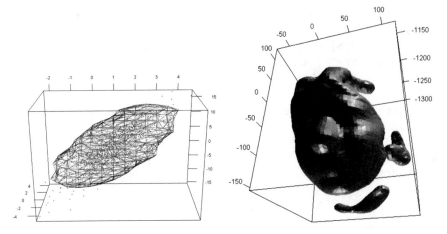

Figure 4.12 Three-dimensional data points and the high-density region; they are given by Program (4 - 20) (left). Rendering of three-dimensional data points; this is given by Program (4 - 21).

```
  val1 <- NULL
  for(ii in 1:nn){
    dis1 <- (x[ii] - xxd)^2 +  (y[ii] - yyd)^2 + (z[ii] - zzd)^2
    s1 <- sort(dis1)
    val1[ii] <- -1/s1[20]
  }
  return(val1)
  }
# (3)
  set.seed(393)
  xxd <<- rnorm(1000, mean = 1, sd = 1)
  yyd <<- rnorm(1000, mean = -1, sd = 1) + xxd
  zzd <<- runif(1000, min = -3, max = 3) + xxd * 2 + yyd * 2
# (4)
  th1 <- -0.4
# (5)
  xx <- seq(from = -3, to = 5, by = 0.5)
  yy <- seq(from = -6, to = 6, by = 0.5)
  zz <- seq(from = -20, to = 20, by = 2)
# (6)
  clear3d(type = "all" )
# (7)
  rgl.light(theta = 40, phi = 40, viewpoint.rel = T,
    ambient = "white", diffuse = "white", specular = "white")
  rgl.light(theta = -95, phi = -30, viewpoint.rel = T,
```

```
      ambient = "white", diffuse = "white", specular = "white")
# (8)
      contour3d(ffb, th1, xx, yy, zz, color = "green",
      engine = "rgl", scale = T, color2 = "green", fill = F,
      distance = 0.05)
# (9)
      points3d(xxd, yyd, zzd, size = 3, color = "magenta")
# (10)
      axes3d()
# (11)
      box3d()
# (12)
      aspect3d(x = 1.5, y = 0.8, z = 1)
}
```

(1) The use of the packages "misc3d" and "rgl" is described.

(2) The function ffb() for drawing a high-density region of data points is defined. The high-density region of data points is defined as the area in which, when squared distances between the data points and other data points are arranged in an ascending order, the inverse number of the 20th value is large.

(3) The simulation data xxd, yyd, and zzd, are produced. Since xxd, yyd, and zzd are used in the function ffb(), <<- is employed.

(4) th1, ia threshold for drawing a constant-height surface of the function ffb(), is given.

(5) The grid values given by xx, yy, and zz are given; the values of the function ffb() are calculated at these points.

(6) The command clear3d(type = "all") deletes all settings concerning the three-dimensional graph.

(7) The command rgl.light() specifies the light settings. Two lights are set in this example.

(8) The command contour3d() draws a constant-height surface. engine = "rgl" indicates that the functions of the package "rgl" are used. The argument fill = F means wire frame objects. The argument distance = 0.05 indicates the distance between the original point and the viewpoint.

(9) The command points3d() draws three-dimensional data points in the three-dimensional graph.

(10) The command axes3d() draws coordinate axes.

(11) The command box3d() illustrates a rectangular solid in the outside area.

(12) The command aspect3d() adjusts the length of the axes of the three-dimensional area.

When engine = "rgl" is specified in contour3d(), several functions of the package "rgl" are available. Furthermore, techniques used in Fig. 4.11 and Fig. 4.12 (left) enable rendering without constructing an implicit function.

Fig. 4.12 (right), given by Program (4 - 21), exemplifies it; the data used here is identical to those in Fig. 4.6 (left)(page 207).

Program (4 - 21)

```
function() {
# (1)
  library(misc3d)
  library(rgl)
# (2)
  ffc <- function(x, y, z) {
    nn <- length(x)
    val1 <- NULL
    for(ii in 1:nn){
      dis1 <- (x[ii] - xxd)^2 + (y[ii] - yyd)^2 +
      (z[ii] - zzd)^2
      s1 <- sort(dis1)
      val1[ii] <- 1/s1[50]
    }
  return(val1)
  }
# (3)
  file1 <- read.table(file = "d:\\GraphicsR\\rgl1.txt",
  header = F)
  xxd <<- file1[,1]
  yyd <<- file1[,2]
  zzd <<- file1[,3]
# (4)
  th1 <- 0.005
# (5)
  xx <- seq(from = -70, to = 120, by = 5)
  yy <- seq(from = -150, to = 130, by = 5)
  zz <- seq(from = -1300, to = -1100, by = 5)
# (6)
  clear3d(type = "all")
# (7)
  rgl.light(theta = 60, phi = 50, viewpoint.rel = T,
    ambient = "white", diffuse = "white", specular = "white")
  rgl.light(theta = -70, phi = -80, viewpoint.rel = T,
    ambient = "white", diffuse = "white", specular = "white")
# (8)
  contour3d(ffc, th1, xx, yy, zz, color = "green",
    engine = "rgl", scale = T, color2 = "green", fill = T,
    distance = 0.05)
# (9)
  view3d(theta = 15, phi = -30)
```

```
# (10)
  axes3d()
# (11)
  box3d()
# (12)
  aspect3d(x = 1, y = 1.2, z = 1)
}
```

(1) The use of the packages "misc3d" and "rgl" is described.

(2) The function `ffc()` for drawing curved surfaces to show high-density regions of data points is defined. The high-density region of data points is defined as the area in which when squared distances between the data points and other data points are arranged in an ascending order, the inverse number of 50th value is large.

(3) The text file, rgl1.txt, is retrieved and named `file1`. Data points stored in `file1` are named `xx` (the x-axis), `yy` (the y-axis), and `zz` (the z-axis).

(4) The values of the function `ffc()` on the constant-height surface are specified as `th1`.

(5) The grid values given by `xx`, `yy`, and `zz` are given; the values of the function `ffc()` are calculated at these points.

(6) The command `clear3d(type = "all")` deletes all settings concerning the three-dimensional graph.

(7) The command `rgl.light()` specifies the light settings. Two lights are set in this example.

(8) The command `contour3d()` draws a constant-height surface. The argument `fill = T` means a constant-height surface.

(9) The command `view3d()` sets the position of the viewpoint for surveying the three-dimensional graph.

(10) The command `axes3d()` draws the coordinate axes.

(11) The command `box3d()` depicts a rectangular solid in the outside area.

(12) The command `aspect3d()` adjusts the length of the axes of the three-dimensional graph.

The package "misc3d" enables surveying the behavior of a function with three predictors from three directions. Fig. 4.13 (left), given by Program (4 - 22), exemplifies it.

```
Program (4 - 22)
function() {
# (1)
  library(misc3d)
# (2)
  ffd <- function(x, y, z) {
    val1 <- exp(-((x + 1)^2 + y^2 + (z + 1)^2)) * 0.2 +
      exp(-((x - 1)^2 + y^2 + z^2)) * 0.3
  return(val1)
  }
```

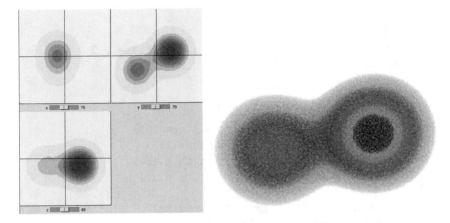

Figure 4.13   Function with three predictors; graph given by Program (4 - 22) (left). Function with three predictors; graph given by Program (4 - 23) (right).

```
# (3)
  xx <- seq(from = -2.5, to = 2.8, length = 150)
  yy <- seq(from = -2.2, to = 3, length = 140)
  zz <- seq(from = -2.8, to = 2.4, length = 130)
  grid1 <- expand.grid(x = xx, y = yy, z = zz)
# (4)
  vv <- array(ffd(grid1$x, grid1$y, grid1$z), c(150, 140, 130))
# (5)
  slices3d(vv, col1 = rev(heat.colors(10)))
}
```

(1) The use of the package "misc3d" is described.

(2) A function with three predictors, `ffd()`, is defined.

(3) The values of grid points where the values of the function `ffd()` are calculated are given as `grid1`.

(4) The values of the function `ffd()` at grid points are brought together as vv in array format.

(5) The command `slices3d()` draws the behavior of the function with three predictors from the three view points.

The package "misc3d" draws the behavior of the function of three predictors using a translucent object. Fig. 4.13 (right), given by Program (4 - 23), exemplifies it.

```
Program (4 - 23)
function() {
# (1)
  library(misc3d)
  library(rgl)
```

```
# (2)
  ffe <- function(x, y, z) {
    val1 <- exp(-((x + 1)^2 + y^2 + (z + 1)^2)) * 0.2 +
      exp(-((x - 1)^2 + y^2 + z^2)) * 0.3
  return(val1)
  }
# (3)
  xx <-seq(from = -2.5, to = 2.8, length = 150)
  yy <-seq(from = -2.2, to = 3, length = 140)
  zz <-seq(from = -2.8, to = 2.4, length = 130)
  grid1 <- expand.grid(x = xx, y = yy, z = zz)
# (4)
  vv <- array(ffe(grid1$x, grid1$y, grid1$z),
   c(length(xx), length(yy), length(zz)))
# (5)
  clear3d(type = "all")
# (6)
  image3d(vv, col = rev(rainbow(5)), jitter = T, alpha =
    c(0.1, 0.2, 0.3, 0.4, 0.5))
}
```

(1) The use of the packages "misc3d" and "rgl" is described.
(2) The function ffe() with three predictors is defined.
(3) The values of grid points where the values of the function ffe() are calculated are given as grid1.
(4) The values of the function ffe() at grid points are brought together as vv in array format.
(5) The command clear3d(type = "all") deletes all settings concerning the three-dimensional graph.
(6) The command image3d() draws a translucent object.

The argument alpha = c(0.1, 0.2, 0.3, 0.4, 0.5) indicates the degree of transparency of the objects.

4.8   PACKAGE "APLPACK"

The package "aplpack" realizes a two-dimensional boxplot, Chernoff faces, a bagplot, a slider, an interactive bootstrap method, an interactive histogram, an interactive density, and an interactive lowess.

A bagplot given by the package "aplpack" illustrates the region that roughly covers data with two predictors and the positions of outliers.

Fig. 4.14 (left), given by Program (4 - 24), exemplifies it.

```
Program (4 - 24)
function() {
# (1)
```

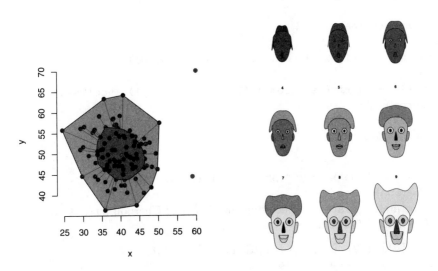

Figure 4.14   Bagplot given by Program (4 - 24) (left). Chernoff faces given by Program (4 - 25) (right).

```
  library(aplpack)
# (2)
  par(mai = c(1, 1, 1, 1), omi = c(0, 0, 0, 0))
# (3)
  set.seed(992)
  data1 <- cbind(rnorm(99, mean = 40, sd = 5),
   rnorm(99, mean = 50, sd = 6))
  data1 <- rbind(data1, c(60, 70))
# (4)
  bagplot(data1, cex = 1, xlab = "x", ylab = "y")
}
```

(1) The use of the package "aplpack" is described.
(2) A graphics area is set.
(3) Simulation data is produced. The data point c(60, 70), which is apparently an outlier, is added.
(4) The command bagplot() draws a bagplot.

For more details about the bagplot, refer to P. J. Rousseeuw, I. Ruts, J. W. Tukey (1999): The bagplot: a bivariate boxplot. The American Statistician, Vol. 53, No. 4, pp.382-387.

Furthermore, Chernoff faces are drawn by the package "aplpack". Fig. 4.14 (right), given by Program (4 - 25), exemplifies it.

Program (4 - 25)

```
function() {
# (1)
  library(aplpack)
# (2)
  par(mai = c(1, 1, 1, 1), omi = c(0, 0, 0, 0))
# (3)
  fmat1 <- matrix(seq(from = 1,to = 72, by = 1), ncol = 8,
  byrow = T)
# (4)
  faces(fmat1)
}
```

(1) The use of the package "aplpack" is described.
(2) A graphics area is set.
(3) The matrix fmat1 for specifying the shapes of Chernoff faces is given.
The matrix fmat1 in this example is:

	[,1]	[,2]	[,3]	[,4]	[,5]	[,6]	[,7]	[,8]
[1,]	1	2	3	4	5	6	7	8
[2,]	9	10	11	12	13	14	15	16
[3,]	17	18	19	20	21	22	23	24
[4,]	25	26	27	28	29	30	31	32
[5,]	33	34	35	36	37	38	39	40
[6,]	41	42	43	44	45	46	47	48
[7,]	49	50	51	52	53	54	55	56
[8,]	57	58	59	60	61	62	63	64
[9,]	65	66	67	68	69	70	71	72

(4) The command faces() draws Chernoff faces.
   For more detail, refer to Chernoff, H. (1973): The use of faces to represent points in k-dimensional space graphically. Journal of the American Statistical Association, vol. 68, pp.361-368.

4.9   PACKAGE "VEGAN"

The package "vegan" treats multivariate analysis used by community ecologists. For more details, refer to the following websites:
http://finzi.psych.upenn.edu/R/library/vignettes/../vegan/doc/
FAQ-vegan.pdf
http://cc.oulu.fi/~jarioksa/softhelp/FAQ-vegan.html
http://ocw.um.es/ciencias/geobotanica/otros-recursos-1/
documentos/vegan.pdf
http://cc.oulu.fi/~jarioksa/opetus/metodi/vegantutor.pdf
http://cc.oulu.fi/~jarioksa/softhelp/vegan/doc/vegan-FAQ.pdf
http://cran.r-project.org/web/packages/vegan/vignettes/
diversity-vegan.pdf

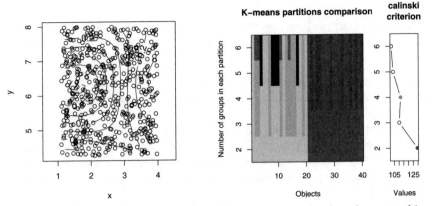

Figure 4.15   Data with two predictors and estimates given by thin-plate smoothing splines; graph drawn by Program (4 - 26) (left). Clustering by the k-means method; graph drawn by Program (4 - 27) (right).

http://cran.univ-lyon1.fr/web/packages/vegan/vignettes/
intro-vegan.pdf
http://ocw.um.es/ciencias/geobotanica/otros-recursos-1/
documentos/vegantutorial.pdf

Data with two predictors and estimates given by thin-plate smoothing splines are drawn by the package "vegan". Fig. 4.15 (left), given by Program (4 - 26), exemplifies it.

Program (4 - 26)

```
function() {
# (1)
  library(vegan)
# (2)
  fun1 <- function(x, y){
    zz <- sin(x) * cos(y)
  return(zz)
  }
# (3)
  par(mai = c(1, 1, 1, 1), omi = c(0, 0, 0, 0))
# (4)
  nd <- 500
  set.seed(921)
  xx1 <- runif(nd, min = 1.1, max = 4)
  xx2 <- runif(nd, min = 4.25, max = 8)
  ww <- seq(from = 1, to = nd, by = 1)
  yy <- fun1(xx1, xx2) + rnorm(nd, mean = 0, sd = 0.3)
  xx <- cbind(xx1, xx2)
```

```
# (5)
  ordisurf(xx, yy, col = "blue", labcex = 1.2, pch = 1,
    xlab = "x", ylab = "y", main = "")
}
```

(1) The use of the package "vegan" is described.
(2) The function fun1() for producing simulation data is defined.
(3) A graphics area is set.
(4) Simulation data is produced. ww gives the weight of each element of data.
(5) The command ordisurf() smooths the data by thin-plate smoothing splines to illustrate the estimates. The package "mgcv" is used to obtain estimates by thin-plate smoothing splines. A smoothing parameter is optimized using *GCV*.

Furthermore, the package "vegan" illustrates the clustering by the k-means method. Fig. 4.15 (right), given by Program (4 - 27), is an example.

```
Program (4 - 27)
function() {
# (1)
  library(vegan)
# (2)
  par(mai = c(1, 1, 1, 1), omi = c(0, 0, 0, 0))
# (3)
  set.seed(126)
  xx1 <- runif(10, min = 1, max = 2)
  xx2 <- runif(10, min = 1, max = 2)
  xx1 <- c(xx1, runif(10, min = 2, max = 3.1))
  xx2 <- c(xx2, runif(10, min = 2.1, max = 2.6))
  xx <- cbind(xx1, xx2)
# (4)
  result1 <- cascadeKM(xx, inf.gr = 2, sup.gr = 6, iter = 30,
    criterion = 'calinski')
  plot(result1)
}
```

(1) The use of the package "vegan" is described.
(2) A graphics area is set.
(3) Simulation data is produced.
(4) The command cascadeKM() carries out clustering using the k-means method. plot() illustrates the result.

4.10 PACKAGE "TRIPACK"

Moreover, the package "tripack" realizes Voronoi tessellation to draw the result. Fig. 4.16 (left), given by Program (4 - 28), is an example.

Program (4 - 28)

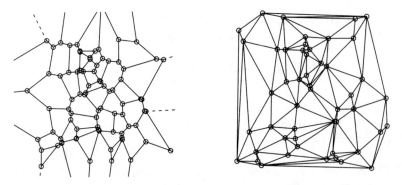

Figure 4.16   Voronoi tessellation yielded by Program (4 - 28) (left). Delaunay triangulation yielded by Program (4 - 29) (right).

```
function() {
# (1)
  library(tripack)
# (2)
  par(mai = c(1, 1, 1, 1), omi = c(0, 0, 0, 0))
# (3)
  set.seed(531)
  xx1 <- runif(50, min = 5, max = 10)
  xx2 <- runif(50, min = 5, max = 10)
# (4)
  voro1 <- voronoi.mosaic(xx1, xx2, duplicate = "remove")
  plot(voro1, main="")
}
```

(1) The use of the package "tripack" is described.
(2) A graphics area is set.
(3) Simulation data is produced.
(4) The command `voronoi.mosaic()` carries out Voronoi tessellation to output the result as `voro1`. The command `plot()` shows a graphic image using `voro1`.

Furthermore, the package "tripack" shows the result of Delaunay triangulation. Fig. 4.16 (right), given by Program (4 - 29), exemplifies it.

Program (4 - 29)
```
function() {
# (1)
  library(tripack)
# (2)
  par(mai = c(1, 1, 1, 1), omi = c(0, 0, 0, 0))
# (3)
  set.seed(531)
```

```
  xx1 <- runif(50, min = 5, max = 10)
  xx2 <- runif(50, min = 5, max = 10)
# (4)
  tri1 <- tri.mesh(xx1, xx2, duplicate = "remove")
  plot.tri(tri1)
}
```

(1) The use of the package "tripack" is described.
(2) A graphics area is set.
(3) Simulation data is produced.
(4) The command `tri.mesh()` carries out Delaunay triangulation to output the results as `tri1`. The command `plot()` shows a graphic image using `tri1`.

4.11 PACKAGE "ADE4"

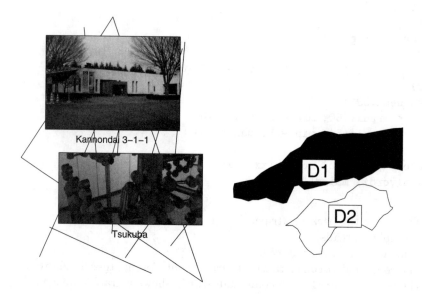

Figure 4.17 Images and straight lines, given by Program (4 - 30) (left). Map given by Program (4 - 31) (right).

The package "ade4" aims to treat ecological and environmental data. "ade4" stands for "data analysis functions to analyze ecological and environmental data in the framework of Euclidean exploratory method." This package enables statistical calculation and graphs. Moreover, this package contains data, redaction, and bibliographic references on ecology and environment. For more details, refer to the following:
http://pbil.univ-lyon1.fr/ade4/home.php?lang=eng

http://cran.md.tsukuba.ac.jp/web/packages/ade4/ade4.pdf

The command citation("ade4") outputs:

Dray, S. and Dufour, A.B. (2007): The ade4 package: implementing the duality diagram for ecologists. Journal of Statistical Software. 22(4): 1-20.

Chessel, D., Dufour, A.B., and Thioulouse, J. (2004): The ade4 package-I-One-table methods. R News. 4: 5-10.

Dray, S., Dufour, A.B., and Chessel, D. (2007): The ade4 package-II: Two-table and K-table methods. R News. 7(2): 47-52.

To cite ade4 in publications, use the above references.

Fig. 4.17 (left), given by Program (4 - 30), shows an example of the graphs that the package "ade4" constructs.

Program (4 - 30)
```
function() {
# (1)
  library(ade4)
  library(pixmap)
# (2)
  par(omi = c(0, 0, 0, 0))
# (3)
  w1 <- data.frame(id = rep("a", 3), x = c(100, 50, 0),
   y = c(-25, 115, 40))
  area.plot(w1)
# (4)
  ppm1 <- read.pnm("d:\\GraphicsR\\ade4_1aeng.ppm")
  ppm2 <- read.pnm("d:\\GraphicsR\\ade4_1beng.ppm")
# (5)
  logo1 <- list(ppm1, ppm2)
  pos1 <- data.frame(x = c(60, 50), y = c(24, 80))
# (6)
  set.seed(360)
  xx1 <- runif(10, min = 0, max = 100)
  yy1 <- runif(10, min = -25, max = 115)
  xx2 <- runif(10, min = 0, max = 100)
  yy2 <- runif(10, min = -25, max = 115)
  data1 <- data.frame(x1 = xx1, y1 = yy1, x2 = xx2, y2 = yy2)
# (7)
  s.logo(pos1, logo1, add.plot = T, clogo = c(0.8, 0.8, 0.8),
   contour = data1)
# (8)
  text(50, 50, "Kannondai 3-1-1", cex = 1.2)
  text(60,0, "Tsukuba", cex = 1.2)
}
```

(1) The use of the packages "ade4" and "pixmap" is described.
(2) A graphics area is set.

(3) The command **area.plot()** holds the area that three or more points indicate; these points are specified by x = and y =. These points are connected by straight lines.

(4) Two image files in a ppm format are retrieved and named **ppm1** and **ppm2**.

(5) The image data **ppm1** and **ppm2** are stored as **logo1** in list format. The positions where these two images are illustrated is set as **pos1**.

(6) The coordinates of the ends of random straight lines to be drawn in the background are calculated and saved in **data1**.

(7) The command **s.logo()** draws two images and random straight lines. The argument **clogo** = indicates the size of the images. The argument **contour** = specifies the coorinates of the ends of the random straight lines in the background.

Furthermore, the package "ade4" enables users to draw outlines of regions using data of coordinates and to write the names of the regions. Fig. 4.17 (right), given by Program (4 - 31), exemplifies it.

Program (4 - 31)

```
function() {
# (1)
  library(ade4)
# (2)
  par(omi = c(1, 1, 1, 1))
# (3)
  map1 <- read.csv(file = "d:\\GraphicsR\\ade4_2.csv",
    header = F)
# (4)
  nd <- length(map1[,1])
# (5)
  xxd <- map1[,1]
  yyd <- map1[,2]
  xx <- xxd[xxd > 0]
  yy <- yyd[yyd > 0]
# (6)
  kk <- 1
  jj <- 0
  dis1 <- NULL
  disname1 <- "D1"
  for(ii in 1:nd){
    if(yyd[ii] < 0){
      kk <- kk+1
      disname1 <- paste("D", as.character(kk), sep = "")
    }
    else{
      jj <- jj + 1
      dis1[jj] <- disname1
```

```
    }
  }
# (7)
  data1 <- data.frame(district = dis1, x = xx,  y = yy)
# (8)
  val1 <- rev(seq(from = 1, to = length(unique(dis1)), by = 1))
# (9)
  area.plot(data1, val = val1, cleg = 0, clabel = 3)
}
```

(1) The use of the package "ade4" is described.
(2) A graphics area is set.
(3) The text data (ade4_2.csv) is retrieved and saved as map1. The text data (ade4_2.csv) stores the degrees of latitude and longitude of the boundaries of the Chugoku and Shikoku areas in Japan. All of the degrees of latitude and longitude are positive. The values between the the data of the two areas are negative.
(4) The number of data points (map1) is represented as nd. nd stands for the number of data points representing the coastline of the two areas plus one (a negative number between the two areas).
(5) The data in map1 is divided into xxd (coordinates of x-axis) and yyd (coordinates of y-axis). A negative number is omitted from xxd and stored as xx. A negative number is omitted from yyd and stored as yy.
(6) dis1 is constructed to indicate the areas to which each data point of the coordinates corresponds. The name of the first area is D1, and that of the second area is D2.
(7) The data of the coordinates of the areas (dis1) is put together in a data frame (data1).
(8) val1 for numbering each area is constructed.
(9) The command area.plot() draws the shapes of the areas and writes their names. The argument cleg = 0 indicates that a legend is not added. The argument clabel = 3 indicates the size of the names.

4.12 PACKAGE "VIOPLOT"

The package "vioplot" allows users to draw violin plots. A violin plot is slightly different from a boxplot. Fig. 4.18 (left), Program (4 - 32), is an example.

```
Program (4 - 32)
function() {
# (1)
  library(vioplot)
# (2)
  par(mai = c(1, 1, 1, 1), omi = c(0, 0, 0, 0))
```

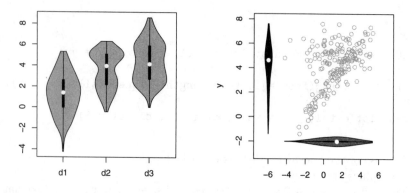

Figure 4.18   Violin plot given by Program (4 - 32) (left). Scatter plot and violin plot given by Program (4 - 33) (right).

```
# (3)
  set.seed(283)
  d1 <-rnorm(200, mean = 1, sd = 2)
  d2 <-c(rnorm(100, mean = 2, sd = 1), rnorm(100, mean = 5,
  sd = 0.5))
  d3 <-c(rnorm(100, mean = 3, sd = 1), rnorm(100, mean = 6,
  sd = 1))
# (4)
  vioplot(d1, d2, d3, names = c("d1", "d2", "d3"), col = "green")
}
```

(1) The use of the package "vioplot" is described.
(2) A graphics area is set.
(3) Simulation data is produced.
(4) The command `vioplot()` draws a violin plot. For more details about the violin plot, refer to the following refrerence:
Hintze, J. L. and R. D. Nelson (1998): Violin plots: a box plot-density trace synergism. The American Statistician, Vol.52, No.2, pp.181-184.

A violin plot can be added to a graph of a two-variable distribution. Fig. 4.18 (right), given by Program (4 - 33), exemplifies it.

```
Program (4 - 33)
function() {
# (1)
  library(vioplot)
# (2)
  par(mai = c(1, 1, 1, 1), omi = c(0, 0, 0, 0))
# (3)
  set.seed(283)
```

```
  xx <-rnorm(200, mean = 1, sd = 2)
  yy <-c(xx[1:100] + rnorm(100, mean = 2, sd = 0.5),
   rnorm(100, mean = 5, sd = 1))
# (4)
  plot(xx, yy, xlim = c(-7, 7), ylim = c(-3, 8), xlab = "x",
   ylab = "y", col = "green")
# (5)
  vioplot(xx, col = "magenta", horizontal = T, at = -2, add = T)
  vioplot(yy, col = "blue", horizontal = F, at = -6, add = T)
}
```

(1) The use of the package "vioplot" is described.
(2) A graphics area is set.
(3) Simulation data is produced.
(4) Data points with two variables are plotted in a scatter plot.
(5) The command `vioplot()` draws a violin plot along the x-axis to show the distribution of `xx`. The command `vioplot()` draws a violin plot along the y-axis to show the distribution of `yy`.

4.13 PACKAGE "PLOTRIX"

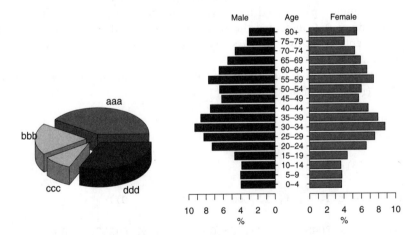

Figure 4.19   Three-dimensional pie chart, given by Program (4 - 34) (left). Population pyramid given by Program (4 - 35) (right).

The package "plotrix" enables various graphics such as displaying text on a circular arc, classifying wind direction and speed records, displaying a centipede plot, plotting values on a 24-hour "clockface", displaying Gantt charts, plotting a multiple histogram as a barplot, and plotting values on a circular grid of 0 to 360 degrees. For more details, refer to `http://cran.r-project.org/web/packages/plotrix/plotrix.pdf`.

For example, Fig. 4.19 (left), given by Program (4 - 34), gives a three-dimensional pie chart.

```
Program (4 - 34)
function() {
# (1)
  library(plotrix)
# (2)
  par(mai = c(1, 1, 1, 1), omi = c(0, 0, 0, 0))
# (3)
  val1 <- c(9, 4.2, 2, 7)
# (4)
  labels1 <- c("aaa", "bbb", "ccc", "ddd")
# (5)
  pie3D(val1, radius = 0.9, labels = labels1, explode = 0.15)
}
```

(1) The use of the package "plotrix" is described.
(2) A graphics area is set.
(3) The values to be used in a graph are set as `val1`.
(4) The labels for attaching the graph are specified as `labels1`.
(5) The command `pie3D()` draws a three-dimensional pie chart. The argument `radius = 0.9` indicates the radius of the three-dimensional pie chart. The argument `explode = 0.15` specifies the size of intervals of the three-dimensional pie chart.

Furthermore, the package "plotrix" realizes a population pyramid such as Fig. 4.19 (right) given by Program (4 - 35).

```
Program (4 - 35)
function() {
# (1)
  library(plotrix)
# (2)
  par(mai = c(1, 1, 1, 1), omi = c(0, 0, 0, 0))
# (3)
  male <- c(243648, 246639, 238695, 288446, 449576, 508302,
   573146, 529690, 460461, 378960, 393422, 472951, 396362,
   334449, 281675, 195032, 179859)
  male <- male/sum(male) * 100
  female <- c(233044, 234743, 227898, 274522, 410166, 472928,
   548543, 496326, 424685, 357696, 376632, 465718, 417060,
   371495, 330725, 256325, 345967)
  female <- female/sum(female) * 100
# (4)
  agelabels <- c("0-4", "5-9", "10-14", "15-19", "20-24",
```

```
    "25-29", "30-34", "35-39", "40-44", "45-49", "50-54",
    "55-59", "60-64", "65-69", "70-74", "75-79","80+")
# (5)
  pyramid.plot(matrix(male, ncol = 1), matrix(female, ncol = 1),
    labels = agelabels, lxcol = "blue", rxcol = "red", gap = 2,
    labelcex = 0.9)
}
```

(1) The use of the package "plotrix" is described.

(2) A graphics area is set.

(3) The male population by age is stored in `male`, and the female population by age is stored in `female`. The values of `male` and `female` are normalized; the sum of the values of `male`, as well as of `female`, is 100. This population data is that of Tokyo in 2005 (`http://www.toukei.metro.tokyo.jp/tnenkan/2005/tn05qyti0510b.htm`).

(4) The labels for indicating age intervals are stored in `agelabels`.

(5) The command `pyramid.plot()` draws a population pyramid. The argument `lxcol` = sets the color of the bars showing the male population. The argument `rxcol` = sets the color of the bars showing the female population. The argument `gap` = 2 indicates the width of the space showing the age intervals. The argument `labelcex` = 0.9 sets the size of the letters showing the age intervals.

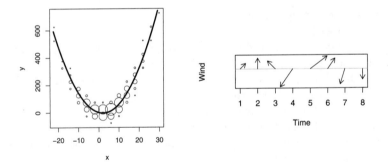

Figure 4.20   Density of data and regression equation given by Program (4 - 36) (left). Time series of wind speed and wind force illustrated by Program (4 - 37) (right).

The package "plotrix" contains the function of illustrating accumulated data clearly. For example, in Fig. 4.19 (left), the areas of circles show the numbers of neighboring data points near the circles. The curve is given by quadratic regression. This graph is yielded by Program (4 - 36).

Program (4 - 36)

```
function() {
# (1)
  library(plotrix)
# (2)
  par(mai = c(1, 1, 1, 1), omi = c(0, 0, 0, 0))
# (3)
  set.seed(435)
  nd <- 1000
  xx <- rnorm(nd, mean = 4, sd = 8)
  yy <- (xx - 2)^2 - 1 + rnorm(nd, mean = 0, sd = 20)
  data1 <- data.frame(x = xx, y = yy)
# (4)
  lm1 <- lm(y~poly(x, degree=2), data = data1)
  ex <- seq(from = min(xx), to = max(xx), length = 100)
  data2 <- data.frame(x = ex)
  ey <- predict(lm1, data2)
# (5)
  xx <- (floor(xx * 0.25) + ceiling(xx * 0.25)) * 0.5 * 4
  yy <- (floor(yy * 0.02) + ceiling(yy * 0.02)) * 0.5 * 50
# (6)
  sizeplot(xx, yy, scale = 0.3, col = "blue", xlab = "x",
  ylab = "y")
  lines(ex, ey, lwd = 3)
}
```

(1) The use of the package "plotrix" is described.
(2) A graphics area is set.
(3) A total of 1000 simulation data points are produced and named xx and yy. xx and yy are combined in data1.
(4) The command lm() carries out quadratic regression using data1. The estimates by the quadratic regression are stored in ey.
(5) The values saved in xx and yy are discretized.
(6) The command sizeplot() represents the number of data in each range as an area of a circle. The command lines() draws a line representing estimates of the quadratic regression.
When the number of data points is one, a circle is drawn.

The package "plotrix" allows users to draw graphs to show the time series of the directions and strengths of the wind. Fig. 4.19 (right), illustrated by Program (4 - 37), exemplifies it.

```
Program (4 - 37)
function() {
# (1)
  library(plotrix)
# (2)
  par(mai = c(1, 1, 1, 1), omi = c(0, 0, 0, 0))
```

```
# (3)
  xx <- c(1.5, 2, 2.1, 4.1, 4, 2.4, 3, 2)
  th1 <- c(0.25, 0.5, 0.75, 1.3, 0.2, 0.3, 1.4, 1.5) * pi
# (4)
  feather.plot(xx * 0.3, theta = th1, xlab = "Time", ylab = "Wind")
}
```

(1) The use of the package "plotrix" is described.

(2) A graphics area is set.

(3) The strength of the wind is stored in xx. The direction of the wind is stored in th1.

(4) The command feather.plot() draws time series of directions and strengths of wind.

4.14 PACKAGE "RWORLDMAP"

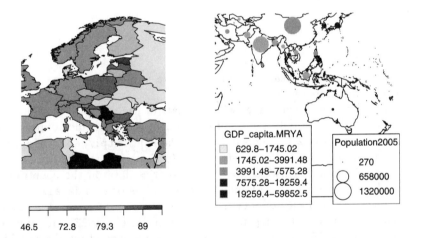

Figure 4.21   Map of Environmental Performance Index (EPI) in Europe, given by Program (4 - 38) (left). Population (areas of circles) and GDP per capita (colors of circles) of countries in Asia and Oceania, given by Program (4 - 39) (right).

The package "rworldmap" draws maps of diverse area in the world. Countries are color-coded on the basis of data. For more details, refer to
http://cran.r-project.org/web/packages/rworldmap/rworldmap.pdf
http://cran.r-project.org/web/packages/rworldmap/vignettes/rworldmapFAQ.pdf

Fig. 4.21 (left) give by Program (4 - 38) shows an example.

Program (4 - 38)

```
function() {
# (1)
  library(rworldmap)
# (2)
  par(mai = c(0.7, 1, 0.7, 1), omi = c(0.7, 0.1, 0.7, 0.1))
# (3)
  data("countryExData", package = "rworldmap")
# (4)
  data1 <- joinCountryData2Map(countryExData, joinCode = "ISO3",
   nameJoinColumn = "ISO3V10", projection = "none")
# (5)
  mapParams <- mapCountryData(data1, nameColumnToPlot =
   "AGRICULTURE", xlim = c(10,25), ylim = c(30,70),
   addLegend = F, mapTitle = "", colourPalette =
   "terrain", missingCountryCol = "blue", borderCol = "black")
# (6)
  do.call(addMapLegend, c(mapParams, legendLabels = "all",
   legendWidth = 0.8))
}
```

(1) The use of the package "rworldmap" is described.

(2) A graphics area is set.

(3) The data countryExData, which is part of the package "rworldmap", is retrieved. The data indicates the Environmental Performance Indices (EPIs) of 149 countries in 2008. This data is available at http://epi.yale.edu/.

(4) The command joinCountryData2Map() associates the data in hand (countryExData in this example) with contries data on the countries contained in the package "rworldmap"; the result is stored in data1.

(5) The command mapCountryData() draws a map to color code it on the ground of data1. The map uses the data of AGRICULTURE. The arguments xlim = c(10, 25), ylim = c(30, 70) extract a region from the world map. The argument colourPalette = "terrain" indicates the colors to be used for the color coding; colorPalette = does not work here. Along with terrain, white2Black, black2White, palette, heat, topo, terrain, rainbow,negpos 8, and negpos9 are available. If colourPalette = is not specifed, heat is set. The argumemt missingCountryCol = "blue" indicates that countries lacking data are painted with blue. The command mapParams <- stores positions in the map drawn by mapCountryData() in mapParams; mapParams is used in (6).

(6) The command do.call() executes addMapLegend(). It adds a legend to the map constructed in (5).

The package "rworldmap" illustrates the maps of various areas in the world and show the data of each country by adding circles or other symbols on the position of the country. For example, Fig. 4.21 (right), depicted by Program (4 - 39), is an example.

Program (4 - 39)
```
function() {
# (1)
  library(rworldmap)
# (2)
  par(mai = c(0.1, 1, 0.1, 1), omi = c(0.1, 0.1, 0.1, 0.1))
# (3)
  data("countryExData", package = "rworldmap")
# (4)
  data2 <- joinCountryData2Map(countryExData, joinCode = "ISO3",
    nameJoinColumn = "ISO3V10")
# (5)
  mapBubbles(data2, nameZSize = "Population2005", nameZColour =
    "GDP_capita.MRYA", colourPalette = 'topo', numCats = 5,
    catMethod = "quantiles", mapRegion = "oceania",
    borderCol = "black")
}
```

(1) The use of the package "rworldmap" is described.
(2) A graphics area is set.
(3) The data `countryExData`, which is part of the package "rworldmap", is retrieved.
(4) The command `joinCountryData2Map()` associates the data in hand (`countryExData` in this example) with contries data of countries contained in the package "rworldmap"; the result is stored in `data1`.
(5) The command `mapBubbles()` draws a map and add circles on the ground of `data2`. The argument `nameZSize = "Population2005"` indicates that each size of a circle means the population of the corresponding country in 2005. The argument `nameZColour = "GDP_capita.MRYA"` specifies that the color of a circle represents the value of GDP per capita of each country. "MRYA" means "most recent year available." The argument `numCats = 5` indicates that the color of a circle is color coded using five colors. The argument `mapRegion = "oceania"` indicates that a map of Oceania is drawn.

EXERCISES

4.1   Display the logo of R by adding the following three lines. If the setting satisfies some conditions, `package = "rimage"`, an argument of `data()`, is not needed.

```
  library(rimage)
  data(logo, package = "rimage")
  plot(logo)
```

4.2 Confirm that the 95 percent confidence intervals of the population means depicted in Fig. 4.2 (right)(Program (4 - 3)(page 196)) are given by Eq. (4.1).

4.3 Produce other color samples by modifying Program (4 - 5)(page 199).

4.4 Modify the contents of `opts()` in (7) in Program (4 - 7)(page 202) to adjust the appearance of the graph differently.

4.5 Modify the command below in part (3) of Program (4 - 9)(page 204) to observe the change of the graph.

```
if(s1[20] <= 5) val1 <- 1
```

4.6 GavabDB contains many face data points. Choose one of them to draw graphs in a similar manner to Fig. 4.6 (left)(Program (4 - 10)(page 207)).

4.7 Add the following three lines to Program (4 - 12)(page 209) to view the three-dimensional objects from a rotating and ascending viewpoint. If your PC is a high-speed one, the movement of the viewpoint will be too fast. Meaningless procedures make the movement of the viewpoint slower.

```
for(ii in 0:30) {
    rgl.viewpoint(theta = 60 - ii * 10, phi = 2 + ii * 2)
}
```

4.8 Replace (2)(3)(4) of Program (4 - 17)(page 222) with:

```
# (2)'
  fx1 <- function(u, v){
    val1 <- (2 + cos(u)) * cos(v)
  return(val1)
  }
# (3)'
  fy1 <- function(u, v){
    val1 <- sin(u)
  return(val1)
  }
# (4)'
  fz1 <- function(u, v){
    val1 <- (2 + cos(u)) * sin(v)
  return(val1)
  }
```

4.9 Replace the three-dimensional data employed in Program (4 - 19)(page 224) with other simulation data or real data to draw similar graphs.

4.10 Program (4 - 20)(page 226) does not consider that the dispersions of data along the x-axis, y-axis, and z-axis vary. Hence, Fig. 4.12 (left)

exaggerates the distance along the z-axis. Modify the part of (2) in Program
(4 - 20) to ameliorate the situation.

4.11 To acquire a distribution obtained by combining two normal distributions, replace part (2) in Program (4 - 23)(page 231) with:

```
# (2)'
  ffe <- function(x, y, z) {
    val1 <- dnorm(x, mean = 0.8, sd = 1) *
    dnorm(y, mean = 0.5, sd = 0.5) * dnorm(z,
    mean = -0.2, sd = 1) + dnorm(x, mean = 0, sd = 0.5) *
    dnorm(y, mean = 0, sd = 0.8) * dnorm(z, mean = -0.5,
    sd = 0.5)
  return(val1)
  }
```

4.12 To draw a boxplot obtained by projecting data with two variables in a certain direction, replace part (4) of Program (4 - 24)(page 232) with:

```
# (4)'
  plot(data1, xlab = "x", ylab = "y", pch = 4)
  boxplot2D(data1, box.shift = 40, angle = 0)
  boxplot2D(data1, box.shift = 0, angle = -pi * 0.25)
```

4.13 The commands vegdist() and spantree() draw a minimum spanning tree, which is a technique for combining similar data. To realize it, replace part (4) of Program (4 - 27)(page 236) with:

```
# (4)'
  dis1 <- vegdist(xx)
  tr1 <- spantree(dis1)
  plot(tr1, type = "t", labels = as.character(seq(from = 1,
  to = 20, by = 1)))
```

4.14 Add the following four lines at the end of Program (4 - 28)(page 236). These four lines write the values of the areas of 10 segments given by Voronoi tessellation at the centers of the segments; this is a function of cells().

```
  vcells <- cells(voro1)
  for(ii in 1:10){
    text(vcells[[ii]]$center[1], vcells[[ii]]$center[2],
      signif(vcells[[ii]]$area, digit = 2))
  }
```

4.15 To draw the coastline of the Shikoku area in an ordinary scatter plot, replace part (9) of Program (4 - 31)(page 240) with:

```
# (9)'
  area2 <- area2poly(data1)[[2]]
  plot(area2)
  lines(area2)
```

4.16   To determine the relationship between a violin plot and a boxplot, link the following program to the end of Program (4 - 32)(page 241) for adding a boxplot.

```
# (5)
  boxplot(d1, d2, d3, add = T)
```

4.17   To write text in an arc using `arctext()`, add the following program to the end of Program (4 - 34)(page 244).

```
arctext("This shows how pie3D() works", center = c(-0.05, 0.05),
 radius = 0.9, start = NA, middle = pi/2, stretch = 1.5, cex = 1.3)
```

4.18   To show the density of data by numbers using `count.overplot()`, replace part (6) of Program (4 - 36)(page 245) with:

```
# (6)'
  count.overplot(xx, yy, col = "blue", xlab = "x", ylab = "y")
  lines(ex, ey, lwd = 3)
```

CHAPTER 5

APPENDIX

A.1 DIGITAL FILES

To produce image1.txt, an image shot by a digital camera is inverted because
`image()` displays an image upside-down. In addition, if the number of pixels
is too large, `image()` cannot display it. In such a case, the number of pixels
should be reduced. For example, the image is shown on a display of a PC to
be captured by screen-capturing software.

Furthermore, the image file is transformed into a text file consisting of
integers ranging from 0 to 255 to be displayed by `image()` as a black-and-
white image. For this purpose, the free software GIMP2.6, for example, is
used to convert the image file to an image file in PGM (Portable Gray Map)
format. If ASCII is designated at the time of conversion, the resultant image
file is a text file such as:

```
P2
# CREATOR: GIMP PNM Filter Version 1.1
323 389
255
84
```

Guidebook to R Graphics Using Microsoft Windows,
First Edition. By Kunio Takezawa
Copyright © 2012 John Wiley & Sons, Inc.

88
96
95
87
85
87
83
92
85

The rest is omitted.

The third line indicates that the number of pixels of this image file is 323×389. The fourth line shows that the following numbers are integers ranging from 0 to 255. Therefore, if the first four lines of this file are deleted, the resulting text file is a data file consisting of 323×389 integers (pixels).

A.2   FREE SOFTWARE

Several free programs, listed below, are used for constructing this book. I would like to express deep gratitude to the producers of these programs.

1. Images obtained by capturing the display of the personal computer are used in this book. The free programs "Screen Album Ver.1.0" (Author: TAKAO, http://www.clks.jp/) and "QShot" (Author: mira96,

   http://www.vector.co.jp/soft/winnt/art/se470643.html,

   http://mira96.com) were used for this purpose.

2. Movies that will be available at the Wiley website were produced using "Madoroku(DesktopCam400)" (http://soft.noniwa.net/ DesktopCam/) and "AVI Screen Capture 0.2.0.0"

   (http://www.evanolds.com/sc_main.html).

3. In these movies, the position of a mouse and the timing of clicking are shown by "Wink32" (http://kikyoya.cs.hkg.ac.jp/).

4. "GIMP2.6" (http://www.gimp.org/) is used to transform the formats of image files.

A.3   DATA

The data sets below are used for this book. The author is grately for their use.

1. The data for constucting the map of Japan (japan1.txt: Fig. 2.19(page 108)) was produced at "CraftMAP" (Arthor: Shinya Nakamura

   `http://www.craftmap.box-i.net/`); the image data obtained at this web page was binarized. The binarization was carried out using the free software "GIMP2.6". The image file was transformed in PGM (Portable Gray Map) format with the ASCII option and the first four lines of the resultant file were deleted.

2. The data of the population densities in Japan on September 30, 1998 (Fig. 2.19(page 108)) was obtained at

   `http://www.mayors.or.jp/research/jinkou/19990331/3-4.html`.

3. The data of the population pyramid of Japan on October 1, 2006 (pop1.csv: Fig. 2.9(page 87)) was obtained at

   `http://www.stat.go.jp/data/jinsui/2006np/index.htm\#05k18-b`.

4. The data of the shape of a head (rgl1.txt: Fig. 4.6 (left)(page 207)) was downloaded at "GavabDB: face database"

   (`http://gavab.escet.urjc.es/recursos_en.html#GavabDB`) in "Resources of Interest." Details of this data are described in the following article:

   A. B. Moreno, A. Sánchez(2004): GavabDB: a 3D Face Database. Proc. 2nd COST Workshop on Biometrics on the Internet: Fundamentals, Advances and Applications, C. Garcia et al. (eds): Proc. 2nd COST Workshop on Biometrics on the Internet: Fundamentals, Advances and Applications, Ed. Univ. Vigo, pp. 77-82.

   `http://gavab.escet.urjc.es/articulos/GavabDB.pdf`

   `http://www.lsi.upc.edu/~nlp/papers/hernando_securing.pdf`

   rgl1.txt is constructed by extracting data of the positions of a face from cara1_abajo.wrl. The digital file cara1_abajo.wrl is part of face1.zip, which is contained in CD1_faces1-20.zip. The number of data points of rgl1.txt is 12046.

5. The data rgl2.txt used in Fig. 4.6(right)(page 207) is face data with two variables(). This data was produced by extracting part of the data rgl1.txt (Fig. 4.6(left)(page 207) and transfoming it to make the target variable a one-valued function of predictors. That is, after the coordinate transformation of part of the data rgl1.txt used in Fig. 4.6(left) was performed, the randomly selected 5000 data points were interpolated using a thin-plate smoothing spline. As a result, the values of predictors are located at grid points in the data rgl2.txt and the value of the target variable at each grid point is one-valued.

6. The coastline data of the Chugoku and Shikoku areas of Japan (ade4_2.csv: Fig. 4.17(right))(page 238) is constructed at the following website; some

of the data are extracted and "NA"s in the data were replaced with negative values.

Coastline Extractor

`http://rimmer.ngdc.noaa.gov/mgg/coast/getcoast.html`

The Coastline Extractor was created by Rich Signell of the U.S. Geological Survey/Geologic Division/Coastal and Marine Program/Woods Hole Field Center. The extractor pages are currently hosted by the NOAA/National Geophysical Data Center, Marine Geology and Geophysics Division.

Index

.RData, 3, 60
2-dimensional histograms, 195
2-dimensional boxplot, 232

abline(), 186
add =, 37, 136
addMapLegend(), 248
adj =, 25, 26
alpha =, 63, 232
analytical, 124, 125, 128, 133
angle =, 39, 46, 74, 92
apply(), 188, 190
arctext(), 252
area.plot(), 240, 241
argument, 5, 13, 65–67, 177, 211, 249
array, 206
array format, 195, 231, 232
arrow, 74
arrows(), 74, 92, 133
as.numeric(), 147
ASCII, 253, 255
aspect3d(), 217, 221, 228, 230
at =, 89
atpen =, 190
axes =, 59, 60
axes3d(), 211, 217, 221, 228, 230

axis(), 81, 83, 89, 97, 132, 134
azimuth, 102, 179

bagplot, 232, 233
bagplot(), 233
balloon plot, 195, 198, 199
balloonplot(), 199
bandwidth, 72, 185, 186, 188
bar chart, 148, 195
bar plot, 42–44, 59, 61, 83–85, 87, 92,
        110, 112, 132, 180, 190
barplot, 156
barplot, 177
barplot(), 42–45, 61, 84, 85, 89, 91, 92,
        97, 133, 177, 190, 195
beside =, 45
bg =, 37, 110
bin, 68, 69
bin width, 69
black-and-white image, 59, 61
black2White, 248
blue =, 63
bluered(), 200
body(), 128, 137
bold, 35
bold italic, 35

Guidebook to R Graphics Using Microsoft Windows,
First Edition. By Kunio Takezawa
Copyright © 2012 John Wiley & Sons, Inc.

`border =`, 39, 97
boundary, 69
boundary point, 68, 70
`box()`, 97
`box3d()`, 228, 230
boxplot, 35, 38, 61, 75, 76, 132, 135, 203, 241, 251, 252
`boxplot()`, 76, 132, 134
`boxplots =`, 38
`boxwex = 0.5`, 134
break, 142, 160, 170, 174
break point, 115
`breaks =`, 68
`bringToTop()`, 148, 153, 156, 170, 189
`bw =`, 72

`c()`, 124
`cascadeKM()`, 236
`ceiling()`, 76, 178, 180
`cells()`, 251
centipede plot, 243
`cex =`, 25, 76, 94, 110
`cex.axis =`, 94
`cex.labels =`, 94
Chernoff faces, 232–234
chi-squared distribution, 74, 76, 79
circle, 35, 37, 61
`circles =`, 37, 110
circular arc, 243
circular grid, 243
`clabel =`, 241
`clear3d(type = all)`, 209, 211, 213, 216, 218, 221, 223, 228, 230, 232
`cleg =`, 241
click, 141–143, 166, 174, 178, 180, 189
clockface, 243
`clogo =`, 240
`close.screen(all.screen = T)`, 160, 174, 177, 180, 188, 189
clustering, 236
`cm.colors()`, 64
`code =`, 74
`col =`, 30, 61, 64, 110, 124, 141, 166, 177
colatitude, 102, 179
`colfunc4()`, 115
color sample, 195, 199
`colorpanel()`, 199
`colourPalette =`, 248
column, 188
confidence interval, 133, 195–197, 202, 250
console window, 4, 6, 11, 60, 143, 145, 156, 166, 170, 183, 189
constant term, 185

constant-height surface, 223, 224, 226, 230
contour, 100–102, 104, 120, 127, 130, 133
`contour =`, 240
`contour()`, 100, 104, 106, 121, 127, 130, 133
`contour3d()`, 226, 228, 230
Copy as metafile, 18, 62
`cos()`, 99
`count.overplot()`, 252
CRAN, 2
`cube3d()`, 211, 214
cumulative distribution function, 123, 124
`curve()`, 98, 99, 122, 125, 132, 133

`D()`, 125, 128, 130, 133
data file, 254
data frame, 94–96, 169, 177, 196, 197, 204, 206, 221, 241
degrees of freedom, 74
Delaunay triangulation, 237, 238
delta function, 118, 119
`density =`, 39, 46
`density()`, 72, 132
`dev.off(kk)`, 183, 189
`dev.off(which = dev.prev())`, 147
`dev.set(1)`, 59–61, 183
`dev.set(which = dev.next())`, 149, 152, 154, 156, 189
DFFITS, 169
`dffits()`, 169
differentiate, 124, 133
digital camera, 58
digital file, 20–24, 34, 62, 63, 65
digital files, 19
directions and strengths of the wind, 246
discretized, 246
display, xi, 88
`distance =`, 228
`do.call()`, 248
double-click, 158, 174, 175
download, 2
`dpois()`, 137

edge detection, 195
`edges =`, 221
edit screen, 9
editor, 5, 9, 60
`elipse()`, 121
`engine =`, 228
enhanced barplot, 195
enhanced heatmap, 195
Environmental Performance Indices (EPIs), 248

error, 70
error bar, 195
error message, 146, 152, 156, 175, 189
escape key, 141, 170, 177, 188, 189
eval(), 125
exit1(), 147, 148, 152
expand =, 136
explode =, 244
expression form, 128
expression format, 125, 127, 128, 130
expression(), 26, 61

faces(), 234
family =, 21, 62
feather.plot(), 247
female data, 89
fg = red, 37
fig =, 57, 58, 61
figs =, 50
figure area, 11, 48
figure margin, 11, 48, 50, 51, 57, 58, 83
file name, 19, 156
filled.contour(), 101, 133
fin =, 88
floor(), 69, 76, 132
folder, 19, 22, 23, 156
font, 20, 21, 34, 62
font =, 34, 35, 61
for(), 128, 130, 160, 161, 170, 174
forced quit, 146, 149, 150, 152, 156, 189
formula =, 201
free software, 253
free-form surface, 175
frequency, 115, 221
frequency distribution, 90
front, 148, 153, 156, 170, 189
functrion of many variables, 125, 128

Gantt chart, 243
gap =, 245
garaphics window, 183
Gaussian kernel, 72, 185
GavabDB, 250, 255
GCV, 236
geom =, 203
geom = c(point, smooth), 201
ggsave(), 202
GIMP2.6, 253–255
gradient, 128
graphical interface, 140
graphics area, 50, 54, 55
graphics window, xi, 6, 11, 18, 19, 46,
        47, 49, 51, 52, 56, 59–61, 88,
        140–142, 145, 146, 158, 166,
        174, 178, 183, 188, 189
graphics windows, 52, 61
graphics.off(), 6, 146, 189
gray(), 31, 61
grayscale image, 107
green =, 63
grid, 51, 54, 55
group = gg, 203
group.names, 74

hat matrix, 169, 170
heat, 248
heat-transfer function, 116, 118
heat.colors(), 64, 211
height =, 19, 22, 23
heights =, 55, 61, 91
Helvetica, 20, 34, 35
highlight.3d =, 204
hist(), 67–69, 71, 132
histogram, 67, 68, 89, 91, 113, 115, 116,
        132, 133, 221
horiz =, 19, 43, 84

identify(), 143, 189, 190
image file, 253
image matrix, 195
image(), 58, 61, 64, 108, 110, 133, 136
image3d(), 232
imagematrix(), 195
implicit function, 120, 121, 228
inch, 13, 19, 23
inches =, 37
inches =, 110
indefinite integral, 132
integral constant, 132
integral range, 132
integrate(), 132, 133
interactive bootstrap method, 232
interactive density, 232
interactive histogram, 232
interactive image slices, 222
interctive lowess, 232
interpolate, 175, 255
italic, 35

jit =, 74
joinCountryData2Map(), 248, 249
jpeg, 21, 22, 62, 194
jpeg (Joint Photographic Experts Group),
        18
jpeg(), 22, 62

k-means method, 236
kde2d(), 216, 217

kernel =, 72
kernel method, 72
keyboard, 140

labelcex =, 245
labels = ch1, 143
lambda =, 74
las = 1, 62
las =, 43
layer(), 203
layout(), 55, 56, 61, 90, 133
least squares method, 181, 190
least-squares method, 204
left-click, 18
legend, 41, 61, 63, 201
legend(), 41, 42, 61
length =, 74, 92
letters[], 33, 177
levels =, 104
light source, 178–180
limnla =, 177
line =, 82
line chart, 13, 33, 83–85, 132
lines(), 27–30, 33, 60, 61, 72, 124, 132, 142, 160, 161, 246
list format, 74, 80, 240
list(), 74, 80
ljoin =, 39
lm(), 169, 183, 185, 189, 221, 246
local linear regression, 183, 185–189, 191
locations =, 96
locator(), 141, 142, 147, 148, 152, 153, 156, 160, 161, 166, 170, 174, 178, 180, 186, 189
loess, 191
loess(), 191
log(), 135
log =, 42, 61
logarithmic scale, 42, 61, 63
logo of R, 249
lower hinge, 75, 76, 79
lower whisker, 76, 79
lowess, 195
lphi =, 104, 136
lsfit(), 186
ltheta =, 102, 136
lty =, 29, 61
lwd =, 28, 61, 72
lxcol =, 245

mai =, 11, 47, 48, 50, 53, 57, 58, 61, 88
mai =, 87
main =, 41, 62
main program, 149

male data, 89
mapBubbles(), 249
mapRegion =, 249
matrix, 59
matrix format, 195
matrix of scatter plots, 93–95, 133
mean, 74, 122, 123, 133
mean =, 74
median, 75, 76, 79
median(), 75
menu, 2, 3
method = jitter, 74
method =, 160, 201
mfrow =, 47, 52, 61
minimum spanning tree, 251
mirror site, 2
missingCountryCol = blue, 248
mosaic plot, 72
mosaicplot(), 72, 134
mouse, 140–143, 158, 174, 178, 180, 188, 189, 226
MRYA, 249
mtext(), 81–83, 133
multiple histogram, 243
multiplier factor, 110

names =, 76
names(), 169
names.arg =, 43–45, 97
nameZColour =, 249
nameZSize =, 249
natural spline, 160, 161, 165
negpos8, 248
negpos9, 248
Newton-Raphson method, 128, 130
nonparametric regression, 191
normal distribution, 74, 76, 79, 94, 122–124, 133
normal distrution, 121
normal Q-Q plot, 183
notch =, 134
Notepad, 9
notepad, 5
numCats =, 249
numerical integration, 131–133

object, 16, 33, 60
objective variable, 165
oh3d(), 214
omi =, 11, 47–50, 53, 57, 61
one valued function, 255
ones place digit, 66
options(locatorBell = FALSE), 147, 189
opts(), 203, 250

ordisurf(), 236
OS, 1, 20, 21, 34, 35
outer margin, 11, 48, 49
outer(), 100–102, 104, 106, 108, 121,
    127, 130, 211
outlier, 233
overplot(), 196

p =, 197
package, 2, 193
package =, 249
package "ade4", 238–240
package "aplpack", 232
package "assist", 177
package "gplots", 195
package "MASS", 216
package "rimage", 221
package "plotrix", 243
package "rgl", 207
package "rimage", 194
package "rworldmap", 247
package "scatterplot3d", 177, 203
package "tripack", 236
package "vegan", 234, 235
package "vioplot", 241
packege "lattice", 193
packge "ggplot2", 200
pairs(), 94, 95, 133, 135
palette, 248
panel.hist(), 135
par(), 11, 47–50, 52, 53, 57, 58, 60–62,
    87, 133
par(new = T), 56–59, 61, 81, 83–85, 87,
    97, 104, 106, 132, 195
par3d(), 209, 214
parallel chart, 96
parameter, 223
parametric3d(), 223
pch =, 31, 33, 42, 61, 74, 95, 141, 206
pdf, 22–24, 62
pdf (Portable Document Format), 18
pdf(), 23, 24, 62
persp(), 102, 104, 106, 108, 110, 112,
    113, 116, 119, 133, 136, 137,
    191
persp3d(), 209, 221
perspective, 104
perspective plot, 101, 102, 104, 106–108,
    112, 113, 116, 119, 133, 136,
    137, 177–179, 208, 209, 211,
    218, 221
PGM (Portable Gray Map), 253, 255
phi =, 102, 104
pie chart, 45, 61, 86, 87, 132

pie(), 45, 46, 61, 87
pie3D(), 244
pixel, 22, 195, 253, 254
plain, 34, 35
plot(), 13–15, 24, 26, 42, 60–62, 72, 160,
    195, 236–238
plot.lm(), 183, 189
plot3d(), 208, 216
plotmeans(), 197
plural variables, 93, 95, 96
points(), 31, 33, 60, 61, 128, 130, 142,
    166
points3d(), 228
Poisson distribution, 74, 76, 79, 137
poly(), 183, 189
polygon, 39, 61
polygon(), 39, 40, 61
polynomial regression, 183, 189
population density, 108, 111, 112
population mean, 197, 250
population pyramid, 87, 133, 243–245
postscript, 18–21, 62, 202
postscript format, 34
postscript(), 19–21, 62
ppm format, 240
predict(), 169
predictor, 165, 185, 186, 188, 223
preprocessed data, 71, 132
print(), 158
print(Finished), 147, 148, 152
print(par()), 110, 113
probability density function, 72, 121–
    124, 132, 133, 198
probability distribution function, 123
pyramid.plot(), 245

qmesh3d(), 213
qplot(), 201, 203
qt(), 198
quadratic form, 121
quadratic regression, 190, 191, 246
quantile region, 76

r =, 104, 136
R command, 139
R program, 139, 143
radar chart, 95, 96, 133
radius =, 212, 244
rainbow, 248
rainbow(), 64, 104, 209
rchisq(), 74
readline(), 147, 153, 158, 189
realization, 74, 79, 94
rect(), 39, 61

rectangle, 35, 37–39, 61
rectangular solid, 211
recursive, 98, 99, 133
red =, 63
regression flat plane, 204
rendering, 207, 222, 228
repeat, 141, 142, 147, 157, 158, 160, 161, 165, 170, 174, 177, 179, 186, 188
residual, 169, 183
return key, 4–6
return(), 185
rev(), 124
rgl.light(), 209, 211, 213, 216, 218, 221, 223, 228, 230
rgl.spheres(), 211
rgl.surface(), 211, 217
rgl.viewpoint(), 209, 212, 221, 223
right =, 69
right mouse button, 18
rnorm(), 74
rotate3d(), 209, 212, 214
round off, 66
rounding off, 66
rpois(), 74
rxcol =, 245

s.logo(), 240
scale =, 67
scale3d(), 211
scatter plot, 13, 89, 91, 243, 251
scatterplot3d(), 177, 204, 206
screen =, 50
screen(), 50, 51, 160
screen-capturing software, 253
sd =, 74
sea, 112
second derivative, 160
self-similar, 99
seq(), 100, 130
shade = 0.3, 102
shade =, 191
shade3d(), 213
shape =, 203
Sheather and Jones method, 72
side =, 81
simple regression, 166, 190, 221
simulation data, 74, 76, 79, 90, 94–96, 115, 196, 197, 202, 204, 206, 216, 221, 226, 228, 233, 236–238, 242, 243, 246
sin(), 98
size =, 203
slices3d(), 231

slider, 232
smoothing, 177
smoothing parameter, 186, 236
solve(), 130
solve-the-equation method, 72
space =, 177, 180, 190
spantree(), 251
speed record, 243
spheres3d(), 213, 221
splinefun(), 160, 189
split.screen(), 50, 160, 165, 174, 177, 180, 188, 189
square, 37
square matrix, 130
square root, 110
squares =, 37
srt =, 26
ssr(), 177
standard deviation, 122, 133
standard diviation, 74
standardized residual, 183
star, 35, 37, 61
stars =, 37
stars(), 96, 133
stem(), 66, 67, 132
stem-and-leaf display, 66, 67, 132
strip chart, 73, 74, 80, 132
stripchart(), 74, 80, 132
subroutine, 149, 150
substitute(), 125
symbols(), 35, 61, 186

t(), 197
t-distribution, 197, 198
tan(), 135
target value, 186
target variable, 175, 185, 188
tens place digit, 66
terrain, 248
terrain.colors(), 64
text, 5, 19, 24
text file, 253
text(), 25, 26, 34, 60, 61, 110, 113, 122, 136, 148
texture =, 212, 218
thermometer, 35, 37, 61
thermometers =, 37
theta =, 102, 104, 137
thickness of a shade, 102
thin plate smoothing splines, 235
thin-plate smoothing spline, 255
thin-plate smoothing splines, 236
thin-plate spline, 175, 177
three-dimensional contour plot, 221

three-dimensional kernel density estimate, 222
three-dimensional parametric plot, 222
three-dimensional pie chart, 243, 244
`ticksize =`, 72
Times, 21, 35
topo, 248
`topo.colors()`, 64, 213
`trans3d()`, 136
`translate3d()`, 211, 214
transparency, 63
`voronoi.mosaic()`, 238
`type =`, 14, 60, 141, 177, 195, 206, 207

unbiased variance, 197
update, 130
upper hinge, 75, 79
upper probability, 197
upper whisker, 76, 79

variable, 16, 33, 97
variables, 60
`vegdist()`, 251
Venn diagram, 195
`vert =`, 74
`vertices =`, 214
`view3d()`, 208, 217, 230
viewpoint, 178–180
violin plot, 241–243, 252
`vioplot()`, 242, 243
Visual Basic, 139
Visual C++, 139
Voronoi tessellation, 236, 237, 251
`voronoi.mosaic()`, 237

weight, 236
weighted simple regression, 185, 189
`weights =`, 185, 189
`which =`, 170, 183
`white2Black`, 248
`width =`, 19, 22, 23
`widths =`, 55, 61, 90
wind direction, 243
Windows, 1
wire frame, 228
word processor, 18
work image file, 3, 60
`writeLines()`, 147, 148, 152, 158

`xlab =`, 13, 60
`xlim =`, 84, 85, 248

`ylab =`, 13, 60
`ylim =`, 74, 85, 248

zip, 3
`zlim =`, 119